Advanced Topics in Science and Technology in China

Volume 3

Zhejiang University is one of the leading universities in China. In Advanced Topics in Science and Technology in China, Zhejiang University Press and Springer jointly publish monographs by Chinese scholars and professors, as well as invited authors and editors from abroad who are outstanding experts and scholars in their fields. This series will be of interest to researchers, lecturers, and graduate students alike.

Advanced Topics in Science and Technology in China aims to present the latest and most cutting-edge theories, techniques, and methodologies in various research areas in China. It covers all disciplines in the fields of natural science and technology, including but not limited to, computer science, materials science, the life sciences, engineering, environmental sciences, mathematics, and physics.

This book series is indexed by both the **Scopus** and **Compendex** databases.

If you are interested in publishing your book in the series, please contact Violetta Xu (Email: violetta_xuqian@zju.edu.cn) and Mengchu Huang (Email: mengchu.huang@cn.springer.com).

Huiying Zhou · Geng Yang · Baicun Wang ·
Na Dong

Revitalizing Human-Robot Interaction

Smart Manufacturing-Oriented Human Motion Digital Twin

ZHEJIANG UNIVERSITY PRESS
浙江大学出版社

Springer

Huiying Zhou
Dongfang Electric (Hangzhou) Innovation Institute Co., Ltd.
Hangzhou, Zhejiang, China

Baicun Wang
School of Mechanical Engineering
Zhejiang University
Hangzhou, Zhejiang, China

Geng Yang
School of Mechanical Engineering
Zhejiang University
Hangzhou, Zhejiang, China

Na Dong
Dongfang Electric (Hangzhou) Innovation Institute Co., Ltd.
Hangzhou, Zhejiang, China

ISSN 1995-6819 ISSN 1995-6827 (electronic)
Advanced Topics in Science and Technology in China
ISBN 978-981-96-6138-1 ISBN 978-981-96-6139-8 (eBook)
https://doi.org/10.1007/978-981-96-6139-8

Jointly published with Zhejiang University Press Co., Ltd.
The print edition is not for sale in the Chinese mainland. Customers from the Chinese mainland, please order the print book from: Zhejiang University Press Co., Ltd.

This Springer imprint is published by the registered company Springer Nature Singapore Pte Ltd.
The registered company address is: 152 Beach Road, #21-01/04 Gateway East, Singapore 189721, Singapore

Preface

With the development of technology, human-robot interaction has emerged as a highly dynamic field with vast potential for further advancement. This potential is significantly amplified by the key technologies of digitization, networking, and intelligentization. These technologies, in concert, have the potency to redefine the paradigms of interaction between humans and robots. In this progress, it is important to recognize the role humans play. It is not only necessary but crucial to capitalize on the attributes and capabilities of humans to supercharge the efficacy and versatility of the human-robot system.

Firstly, human motion perception and reconstruction constitute the cornerstone of discerning human movements in the human-robot interaction. It necessitates the deployment of state-of-the-art sensor technologies and algorithms. Subsequently, through the application of reconstruction algorithms, based on biomechanics and kinematics of the human body, a faithful digital model of human actions is synthesized. This digital avatar is the basis for subsequent analysis and interaction. Secondly, the comprehensive analysis of human motion assumes a position of importance. On the basis of the statistical inference, machine learning, and biomechanical models, patterns of movement and behavioral intentions can be distilled. This analysis endows robots with the understanding to human actions. The third part focuses on intuitive human-robot teleoperation using Inertial Measurement Unit (IMU)-based motion perception. Using the acceleration, angular velocity, and magnetic field intensity from IMUs, the designed signal processing and control algorithms can transform it into control commands for robots. In the teleoperation, operators can realize their intentions with minimal cognitive overhead, as the robot mirrors human movements. This mode enhances the overall efficiency and efficacy of the interaction. Finally, enhanced human-robot collaboration through motion estimation represents a leap forward. By using motion estimation algorithms, robots can prognosticate the future trajectory of human motion. This predictive prowess enables them to preemptively modulate their actions, effectuating seamless and harmonious coordination. Through the integration of feedback control loops and adaptive strategies, collisions and errors are mitigated, and the safety and reliability of the human-robot ensemble are elevated to new heights.

The overarching ambition of this book is twofold. Primarily, it aims to meticulously dissect and elucidate the key technologies and their sundry applications germane to human-robot interaction, with a particular emphasis in the context of smart manufacturing. This entails a granular exposition of how these technologies can be artfully integrated into industrial processes to supercharge productivity, fortify quality control, and augment flexibility. Secondly, it aspires to function as a lodestar for catalyzing a paradigm shift in the modus operandi of human-machine interaction. By proffering novel concepts, innovative methodologies, and real-world exemplars, the book endeavors to galvanize and inform researchers, practitioners, and industry leaders, propelling the sustainable development of human-machine interaction.

To facilitate the assimilation of complex concepts, the book is embellished with elaborate illustrations that vividly depict technological architectures, motion patterns, and interaction scenarios. These aids serve as conduits for transforming abstract ideas into tangible and easily digestible forms. Complementing these are informative tables that succinctly summarize key data, contrast disparate techniques, and present experimental outcomes in a methodical and organized way. The research algorithms, which form the sine qua non of the technological breakthroughs delineated in the book, are presented in a pellucid and accessible format, accompanied by lucid explanations of their functionality, input-output requisites, and implementation considerations.

This compendious resource is tailored to cater to a heterogeneous audience. Advanced students, eager to plunge into the vanguard of human-robot interaction research, will find it an invaluable pedagogical tool, endowing them with the foundational knowledge and the latest research findings to nourish their academic sojourns. Researchers, questing for novel avenues of exploration and validation of their hypotheses, will reap the benefits of the in-depth analysis and novel ideas proffered. Engineers and developers, tasked with actualizing human-robot interaction systems in real-world applications, will draw inspiration and practical guidance from the detailed technical descriptions and case studies. Finally, entrepreneurs, with an eagle eye on the nascent market opportunities in this domain, will glean insights into the commercial potential and the technological trends that could mold the future of their ventures. In sum, this book is poised to contribute to the advancement of human-robot interaction research and technologies. This monograph embodies the synergistic integration of academic and industrial research efforts. Dr. Huiying Zhou established foundational and innovative methodologies through research work at Zhejiang University and Dongfang Electric (Hangzhou) Innovation Institute Co., Ltd. Professor Geng Yang and Professor Baicun Wang from Zhejiang University spearheaded critical innovations in human-robot interaction and smart manufacturing systems, while Dr. Na Dong at Dongfang Electric (Hangzhou) Innovation Institute Co., Ltd. provided essential industrial insights. We are deeply indebted to our interdisciplinary team for their indispensable contributions. We extend sincere gratitude to Zhejiang University for its institutional support and advanced research infrastructure, and to Dongfang Electric (Hangzhou) Innovation Institute Co., Ltd.

for granting access to manufacturing platforms. This cross-institutional collaboration has demonstrated how academic rigor and industrial pragmatism can mutually reinforce technological advancement.

Hangzhou, China

Huiying Zhou
Geng Yang
Baicun Wang
Na Dong

Contents

Chapter 1
Introduction

Abstract The harmonious coexistence of humans and machines evolves beyond a simple user-machine dynamic into a more interactive, complementary, and supportive ecosystem. This work investigates the human motion digital twin to drive human–robot collaboration in smart manufacturing. The application of human motion digital twin offers substantial support for improving human–robot interaction in this context. On the basis of this trend, this work introduces the research background and significance of digital twin technology in relation to human motion. It then systematically reviews the related work of key technologies for developing human motion digital twin. At last, the main research content and chapter organization of this book are outlined. It emphasizes the in-depth exploration of human motion digital twin within the context of smart manufacturing. This research establishes a roadmap for scalable, human-centric automation-balancing operational efficiency with workforce adaptability in next-gen manufacturing system.

Keywords Digital twin · Smart manufacturing · Human motion · Human interactive behavior · Human motion digital twin

1.1 Background and Motivation

The harmonious coexistence of humans and machines represents a cooperative relationship where both entities mutually enhance and support each other's development. This relationship evolves beyond a simple user-machine dynamic into a more interactive, complementary, and supportive ecosystem. The essence of this harmonious coexistence is to respect and harness human creativity and initiative while using the efficiency and precision of machines, ultimately aiming to achieve the greatest benefit for humanity. This harmonious coexistence is also crucial in smart manufacturing. By integrating advanced information technology, automation, and artificial intelligence, smart manufacturing systems can achieve self-optimization, self-configuration, and self-diagnosis of production processes. It requires not only the efficient and precise

H. Zhou et al., *Revitalizing Human-Robot Interaction*, Advanced Topics in Science and Technology in China 3, https://doi.org/10.1007/978-981-96-6139-8_1

operation of machines but also the innovative thinking and decision-making capabilities of humans to address complex and evolving production needs and market conditions. For instance, in smart factories, robots can adapt based on the skills and experience of human workers, aligning their operations to improve the flexibility and efficiency of the entire production line. This model of human–machine collaboration emphasizes partnership, drives technological advancements and offers innovative solutions to numerous social challenges, which advancing human welfare and societal progress and contributing to overall societal progress [1].

Future advancements in smart manufacturing will depend on the principles of harmonious coexistence to achieve more efficient, cost-effective, and sustainable production methods [2]. In the harmonious coexistence between humans and machines, humans play a vital role. As designers, decision-makers, and participants in this symbiotic system, human initiative and creativity are crucial to its success [3]. Specifically, the significance of humans is reflected in several key areas:

(1) Designers and creators: In smart manufacturing, humans are capable of conceiving and devising machine systems as well as intelligent algorithms. This design process is not arbitrary but is tailored to align with both human requirements and the inherent capabilities of machines. By comprehensively considering factors such as the desired functionality, operational efficiency, and user-friendliness, they are able to engineer technological products that have a profound impact on enhancing various aspects of humans. This includes designing robotic systems for complex tasks in smart factories and developing adaptive algorithms that improve production efficiency and product quality.

(2) Decision-makers and guides: In the context of technological evolution, humans' role extends beyond mere technical innovation to encompass a holistic perspective that takes into account a multitude of factors. As custodians of ethical standards, they consider the moral implications of emerging technologies, ensuring that progress does not come at the expense of human dignity or well-being. In the legal realm, they navigate the complex web of regulations to guarantee compliance and prevent potential legal pitfalls. Culturally, they strive to integrate technological advancements in a manner that respects and preserves the rich tapestry of human values. In the domain of smart manufacturing, this multifaceted responsibility manifests in the pursuit of sustainable development. Humans are tasked with the critical mission of steering the industry towards resource-efficient practices and minimizing environmental degradation. This involves the implementation of strategies such as waste reduction, energy conservation, and the adoption of eco-friendly materials. Concurrently, they must uphold the rights and safety of the workforce, adhering to labor laws and promoting a conducive working environment. By these solutions, they ensure that the benefits of smart manufacturing are equitably distributed.

(3) Operators and recipients: Humans occupy dual roles, being both the operators who interact with machines and the ultimate recipients of the services rendered by these technological marvels. Humans, using their unique cognitive faculties of high creativity and emotional intelligence, are able to concentrate on

complex undertakings such as the conceptualization and design of innovative products and the development of novel processes. In contrast, machines are delegated with the execution of repetitive and labor-intensive work like the precise assembly of components and the management of logistical operations. Moreover, humans bear the responsibility of vigilant supervision over machine operations. This entails continuous monitoring to ensure the seamless functionality of machines, prompt identification and resolution of any operational glitches, and an unceasing effort to enhance machine performance. Such enhancements are calibrated to align with the ever-evolving demands and aspirations of society, which in turn fortifying the collaborative efficacy of the human–machine nexus in the pursuit of manufacturing excellence.

(4) Managers and improvers: As managers, humans oversee machine performance metrics, implementing strategies to bolster production stability and reliability. At the same time, humans are direct beneficiaries of the symbiotic human–machine relationship, enjoying intelligent living and working environments. Smart manufacturing, thus, facilitates the provision of bespoke, high-quality products and services. This harmonious coexistence reaffirms the irreplaceable status of humans.

Maximizing the advantages of human attributes and enhancing capabilities such as perception, cognition, decision-making, learning, and collaboration capabilities within human–machine systems is a critical focus of research in this field. This need has become increasingly pressing within the context of smart manufacturing, which demands a high degree of collaboration between humans and machines to ensure efficient and accurate production processes. Traditional machine-centered interaction paradigms emphasize the interaction between humans and physical systems, but they often fall short in several aspects. They typically lack an understanding of human attributes and states and do not adequately explore the behavior rules and decision-making characteristics of human participants. In smart manufacturing, this traditional approach can result in production processes that fail to fully accommodate the needs and capabilities of workers, potentially impacting efficiency and quality. Additionally, demographic shifts such as an aging population and increasing workforce diversity create higher demands for personnel management and organization. Smart manufacturing requires designing more user-friendly human–machine interfaces and workflows that cater to workers of varying ages and skill levels. This approach will ensure that all workers can effectively engage in the production process.

For example, in industrial settings, welding operations demand stringent quality standards. Workers are frequently exposed to hazardous elements such as lasers, electromagnetic fields, toxic gases, smoke, and dust, which can lead to significant health issues over prolonged periods. Moreover, the welding industry currently encounters challenges including recruitment difficulties, high labor costs, and extended production cycles. In contrast to manual welding, welding robots provide superior stability and quality, with the capability to quantify welding performance through numerical data. They also improve labor productivity, reduce physical strain and operational complexity for workers, and can function in hazardous environments. Additionally, they shorten the preparation time for product redesigns and minimize associated equipment investments [4, 5]. However, current applications centered on welding robots still exhibit certain limitations.

(1) Preparation before welding

 (a) Limitations in workpiece identification and positioning: While welding robots are capable of setting position parameters through programming, they face challenges in identifying and accurately positioning workpieces with complex shapes, irregularities, or significant size deviations [6]. For instance, during the welding of automotive components, some stamped parts may exhibit slight deviations from the design model due to factors such as mold wear and material deformation. While operators can quickly adjust the welding position using visual and tactile feedback, the robot may struggle to detect these errors, leading to inaccurate welding start positions and potentially compromising weld quality.
 (b) Inflexibility in welding parameter adjustment: The welding parameters for robotic systems are typically preset according to established procedures and standard processes. However, prior to actual welding, it is essential to adjust these parameters based on various factors, such as the material of the solder, the thickness of the welded joint, and more. Operators can modify parameters like welding current, voltage, and welding speed on-site, drawing on experience and situations [7]. For example, when welding a workpiece composed of two materials, the operator may adjust heat input to optimize fusion, based on knowledge of the material properties. In contrast, robots without manual intervention are constrained by fixed parameters, which can lead to welding defects.
 (c) Limited fixture adaptability: Fixtures are crucial for securely holding the workpiece during welding to ensure precise positioning. However, robots generally rely on standard fixtures and predefined procedures. If a fixture fails or if a special fixture is needed to accommodate a new workpiece, the robot struggles to adapt as swiftly as an operator. In small-scale manufacturing environments, where small batches and diverse products are common, fixtures may require frequent replacement and adjustment. The robot is often unable to autonomously manage these variable fixture operations, necessitating operator intervention.

(2) Monitoring and strain aspects during welding

 (a) Insufficient real-time monitoring of welding quality: While welding robots can be equipped with sensors to monitor certain parameters, such as arc voltage and current, their ability to assess critical aspects of welding quality, such as weld formation, penetration, and the presence of porosity or slag inclusions, remains limited [8]. Unlike operators, who can visually inspect the weld to identify surface defects such as edge biting or lack of fusion, robots lack the intuitive judgment required to detect and address these issues in real time. Consequently, operators can promptly adjust welding speed or angle in response to visual cues, a flexibility not afforded to robots, thereby potentially compromising the quality of the weld.

(b) Limited ability to respond to emergencies: Welding processes are prone to unexpected disruptions, such as the presence of impurities in welding materials, fluctuations in the power supply, or contamination of the workpiece surface with oil. Experienced operators can swiftly handle the welding process, remove contaminants, or adjust equipment parameters to resolve these issues. In contrast, robots may continue welding according to preprogrammed instructions, even in the presence of these irregularities, which can lead to more severe defects or even damage to the equipment. For example, when a large metal particle contaminates the welding wire, a robot lacks the capability to detect the impurity and pause the welding process as an operator would. This failure to respond can disrupt arc stability and result in defects such as porosity in the weld.

(3) Inspection and repair after welding

(a) Limited weld inspection capability: Upon completion of the welding process, a thorough inspection of the weld is required, such as visual inspection and non-destructive testing. Unlike human operators, robots lack the vision and judgment necessary for performing a detailed visual assessment of the weld, such as evaluating its smoothness or identifying residual spatter. Moreover, for non-destructive testing methods, such as ultrasonic or X-ray testing, robots are unable to independently operate the testing equipment or analyze the resulting data, limiting their capacity to ensure the integrity of the weld [9].

(b) Inability to perform defect repair: When welding defects are identified during the inspection process, repair is typically necessary. Robots generally lack the functionality to perform defect repairs [9]. In contrast, operators can apply appropriate remedial techniques, such as re-welding after surface preparation or utilizing filler materials, depending on the specific type and location of the defect.

The aforementioned welding cases clearly demonstrate that a human-oriented approach is essential for balancing the development needs of human–machine relationships and advancing smart manufacturing. Traditional machine-centric models often reduce human behavior to predefined commands and responses, overlooking the complexity and unpredictability of human actions. In smart manufacturing, worker behavior can be influenced by various factors such as the work environment and task difficulty. Thus, more intelligent human–robot interaction systems are needed to adapt to these changes and integrate human emotions, experience, and intuitive judgment. These factors are essential for solving complex problems and should be incorporated into the interaction systems. Moreover, traditional interaction design often prioritizes system functionality and efficiency while neglecting user experience. In smart manufacturing, enhancing user experience can boost worker motivation and efficiency, then improving overall production performance. Systems designed with traditional paradigms may be powerful but complex to operate and user-unfriendly. They are often inflexible in the face of unknown or non-standard situations and struggle to adapt to variable scenarios or user needs. Machine-centric models may also have operating logic and patterns that do not align with natural human habits, reducing user acceptance and efficiency. Traditional interaction design often employs a "one-size-fits-all" approach, focusing on direct interactions between human users and physical systems. This model ignores the diverse capabilities, needs, and preferences of users. In smart manufacturing, different production roles have distinct requirements for human–machine interaction. For instance, positions requiring high precision may need more sensitive interfaces, while manual labor roles might benefit from simpler, more convenient operational modes. Failure to account for these differences can lead to ineffective or uncomfortable use of technology, especially for elderly users, people with disabilities, or those without technical backgrounds. Therefore, designing inclusive human–robot interaction systems is important. In summary, conducting multi-dimensional and multi-scale digital modeling and analysis of human factors will enhance the focus on human attributes and lead to more efficient and intelligent production processes [3].

Digital twin refers to the creation of a digital model corresponding to a physical object or system in virtual space through digital means [10]. This technology can reflect the status and behavior of physical objects in real-time, leading to effective monitoring and management throughout their lifecycle. Digital twin technology is widely used in industries such as manufacturing, aerospace, construction, medicine, urban planning, and others. In the manufacturing industry, digital twins can help businesses optimize production processes, predict equipment failures, and improve production efficiency. In the aerospace field, digital twins can be used for the lifecycle management of aircraft, from design and manufacturing to operation and maintenance, providing precise data support and decision basis. In urban planning and smart city construction, digital twins enhance the level of city management and the quality of residents' lives through real-time monitoring and analysis of city infrastructure, traffic flow, and environmental data. In the smart manufacturing, this technology enables precise modeling of production equipment and processes within smart factories. It offers lifecycle management, data connectivity, asset management, intelligent simulation, data visualization, and other services, which enhance human-centric approaches and improve human–machine interactions. Digital twins are instrumental in the optimization of production processes, equipment maintenance and management, and worker safety. The COVID-19 pandemic has accelerated the adoption of virtualization and digitization, making remote monitoring and management essential. This situation has underscored the importance of physical interaction and the integration of the physical and digital world, presenting new opportunities for digital twins in smart manufacturing. This integration fosters advancements in unmanned production and remote collaboration within smart factories [11].

The concept of the Human Digital Twin (HDT), which encompasses a comprehensive system, plays a crucial role in human–robot interaction by focusing on human motion intentions and interactions. Wang et al. proposed that HDT is the counterpart of humans reflecting multi-dimensional information, so as to realize the two-way interaction between the physical and the cyber world [3]. It is a model or database of humans, which describes the specific object utilizing key information digitally, such as physical models and physiological models. In addition to the personality and characteristics of human beings, HDT takes the interaction of humans and other resources (e.g., machines, environment) into consideration for depicting humans thoroughly (Fig. 1.1). In smart manufacturing, the interaction between workers and smart devices requires consideration of the human motion digital twin. Therefore, human motion digital twin technology, which involves motion measurement and perception, is essential for creating accurate models of human behavior. This book is dedicated to addressing the challenges and requirements associated with human motion digital twins, with a particular focus on wearable devices. It undertakes an in-depth exploration of key technologies and their corresponding applications, centering on human motion perception and computational strategies based on inertial motion data. By meticulously investigating the behavior characteristics and rules within various interaction scenarios, the book aims to furnish robust theoretical and technical underpinnings for the construction of human-motion digital twin models. Moreover, it endeavors to present innovative solutions for optimizing human–robot interaction within the realm of smart manufacturing. This involves not only enhancing the precision and reliability of human motion capture through wearable devices but also using advanced computational algorithms to process and analyze the acquired data. Consequently, more accurate and efficient digital twin models of human motion can be established, to facilitate seamless integration and interaction between humans and machines in smart manufacturing environments, and ultimately leading to improved productivity and quality in manufacturing processes.

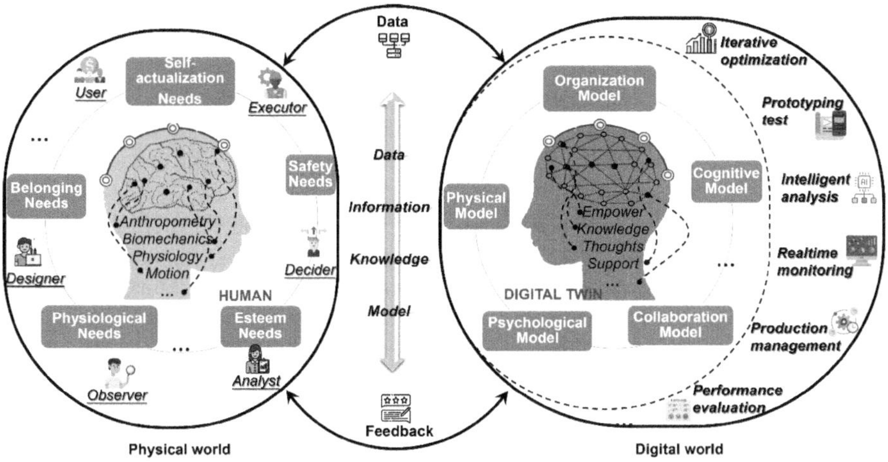

Fig. 1.1 Conceptual framework of human digital twin (HDT) [3]

1.2 Related Work

Human motion digital twin begins by considering the human body's motion dimensions and deepens the application of the digital twin technology of the human body. This section introduces the relevant research status around the key technologies of constructing human motion digital twin.

1.2.1 Research on Human Motion Perception and Analysis

(1) Human posture calculation

Human motion encompasses various forms of movement. Accurate measurement and calculation of human motion are key steps in digitally representing human movements, playing a critical role in understanding and improving human motion performance. By accurately capturing and analyzing human movement data, we can transform complex human dynamics into quantifiable digital information.

Human motion measurement perception digitally presents human physical motion by collecting human motion data. Commonly used human motion capture devices include the following categories: (1) Mechanical motion capture systems, which are technologies that track and measure movement paths through mechanical devices. The mechanical motion capture system primarily consists of multiple joints and rigid rods, in which angle sensors are installed in rotatable joints. These sensors can measure the changes in the rotation angles of the joints. When the device moves, the position and movement trajectory of the endpoint in space can be calculated based on the angle changes measured by the angle sensors and the length of the rods. For instance, the Gypsy Torso motion capture device from Meta Motion company calculates the whole-body skeletal posture from lightweight stereo fisheye cameras attached to a helmet or virtual reality headset [12]. Furthermore, Sharp Warp device also proposes a method to enhance skeletal features in the spine area [13]. (2) Optical motion capture systems, which are devices that use optical principles to track and record the motion of objects, typically composed of multiple high-resolution infrared cameras, markers, data transmission devices, and data processing workstations. These systems record the motion information of an object by placing reflective or luminescent markers on the moving object and then capturing the light from these markers from multiple angles with cameras, such as the Vicon optical system [14, 15] and Qualisys optical system [16, 17] (Fig. 1.2). (3) Image-based motion capture systems, which are devices that utilize visual technology to capture and record the movement of humans or objects. They mainly track and record the motion state of objects by analyzing a series of image frames, with the core being the recognition, positioning, and tracking of moving objects in the video through image processing algorithms, such as the Kinect series devices of Microsoft [18]. (4) An inertial motion capture system, which is a technology that uses inertial sensors to capture and record the movement of objects. By wearing devices integrated with inertial measurement units (IMUs) such as accelerometers, gyroscopes, and magnetometers on key nodes of the moving object, it captures movement data of the target object, including information such as posture and direction of body parts. These data are then sent to the data processing unit via data transmission devices, and after data correction and processing, a three-dimensional model is established, which can move realistically and naturally along with the moving object, for example, Xsens MVN motion capture suit [19, 20] (Fig. 1.2).

While the mechanical motion capture system offers high real-time motion measurement and extensive application scenarios, the wearable rigid mechanical linkage structure affects the wearer's freedom of motion. What's more, its ability to capture the multi-degree-of-freedom joint motion of the human body is poor, which is inconvenient for daily usage. The optical motion capture system has high sampling frequency and high data accuracy but come with high costs, complex installation, and stringent lighting requirements, making them suitable for high-precision laboratory environments [21]. The performance of image-based motion capture systems depends on lighting conditions and background environments. Strong changes in illumination or complex backgrounds can affect the accuracy of marker recognition. At the same time, the cost of setting up and maintaining the system is relatively high, which might be a barrier for some small studios or individual users [22]. The inertial motion capture system is easy to wear, flexible to use, unaffected by space size or coverage, which has a variety of application scenarios. However, inertial sensing devices are prone to interference from linear acceleration of motion and external magnetic field interference [23].

In summary, each type of motion capture system has its advantages and disadvantages, so it is necessary to choose the appropriate motion capture equipment according to different use requirements. In daily applications, human behavior is variable, and the application scenarios are complex. Therefore, the inertial motion capture system is selected as the human motion sensing device in this book.

An IMU typically combines multiple sensors, including a triaxial accelerometer and a triaxial gyroscope, and sometimes may include a triaxial magnetometer. By processing the acceleration data and angular velocity data, the attitude of the sensor can be obtained. However, relying on data from a single sensor can easily lead to increased error. The fusion of multi-source sensor data is helpful to improve the calculation accuracy [24], which is mainly divided into frequency-based methods and modeling-based methods [25].

Wu et al. designed a fast complementary filter with a fixed gain based on a series of magnetic, angular velocity, and gravity sensors [26]. They fused acceleration and angular velocity data and combined the magnetic field strength with the aforementioned estimated gravity values to obtain the target quaternion estimation. Simultaneously, to reduce the significant influence of unknown magnetic field distortions on the magnetometer, they designed a step-by-step filter structure, finding a balance between estimation accuracy and time consumption. However, the fixed complementary gain results in poor real-time motion tracking capability under high dynamic conditions for this filter. Majumder et al. proposed a two-level complementary filter for continuous monitoring of lower limb joints in wearable platforms [27]. This

(a) Optical motion capture system[15]

(b) IMU motion capture suit [19]

(c) A low-cost markerless optical motion capture system for gait analysis [17]

Fig. 1.2 Motion capture systems. **a** Optical motion capture system [15], **b** IMU motion capture suit [19], **c** A low-cost markerless optical motion capture system for gait analysis [17]

two-level complementary filter employs a gradient descent method to estimate the drift of integrated gyroscope data in real time and subsequently applies this method to correct the tilt of the sensor. It estimates the roll and pitch information of two IMUs affixed on either side of the joint by utilizing the acceleration data and angular velocity data from each unit. Then, it integrates the posture information of the two to estimate the angle of the lower limb joint. Although the complementary filter is easy to calculate and realize, it has low robustness to noise interference and is susceptible to output drift from inertial motion data.

The Kalman filter is a method based on modeling analysis, such as predicting the attitude in a short time by using angular velocity, then correcting the error of estimating the attitude by using the magnetic field strength and acceleration of gravity reference point [28]. Baldi et al. proposed a wearable posture estimation system based on IMUs for motion analysis and body tracking [29]. This system used a multiplicative extended Kalman filter, estimated posture through acceleration and angular speed, then reconstructed the position and orientation of the human upper limb. However, due to the absence of integration with magnetic field data, significant errors occur in the posture estimate in the yaw direction. Lee et al. proposes a method of determining time-varying segment-to-joint center vectors to reflect the deformation of the segment-to-joint center vectors and thus to increase the estimation accuracy, in IMMU-based relative position estimation (Fig. 1.3) [30]. Zhu et al. proposed an invariant extended Kalman filter [31]. This filter integrates data returned from the IMU attached to the body with virtual measurements of velocity when standing, enabling it to generate real-time state estimates and fast convergence of human motion estimation errors, even in cases of sensor misplacement and initial state estimate errors. Utilizing the group affine properties of the process model, the stochastic process model is expressed as Brownian motion, and the error value is considered in the estimation of the human leg's motion state. The experimental results show that the algorithm can quickly make the error converge and obtain the real-time state estimation value, even with sensor placement deviations and initial state estimation errors. However, the measurement model adopts the invariant observation form, and its robustness needs to be improved.

Additionally, some researchers have combined human motion characteristics with attitude fusion algorithms. Zhang et al. proposed an adaptive zero-speed update method to tackle accumulated velocity and position errors in pedestrian tracking with inertial sensors of foot [32]. They extracted movement information from chest acceleration and updated the detection threshold for the static phase of gait. Combined with the foot-based inertial navigation algorithm, they obtained foot posture and position across different speeds. Experimental results showed that even when the speed changes rapidly, an accurate tracking trajectory can still be obtained. Zandbergen et al. proposed a drift-free three-dimensional orientation and displacement estimation method based on a single IMU, depending on the quasi-periodic nature of human motion [33]. The method assumes that quasi-periodic motion of the lower limbs occurs within a two-dimensional plane at an approximately constant average speed. Through the application of periodic motion characteristics and the definition of a functional coordinate system with two functional axes, the method reduces attitude drift and achieves posture and displacement estimation of the human lower limbs. Zhou et al. utilized specially designed IMUs with integrated gyroscopes, accelerometers, magnetometers, and processing units [11]. Through data fusion processing, they could obtain information such as acceleration, angular velocity, and magnetic field. They then implemented real-time monitoring of the main parts of the human body

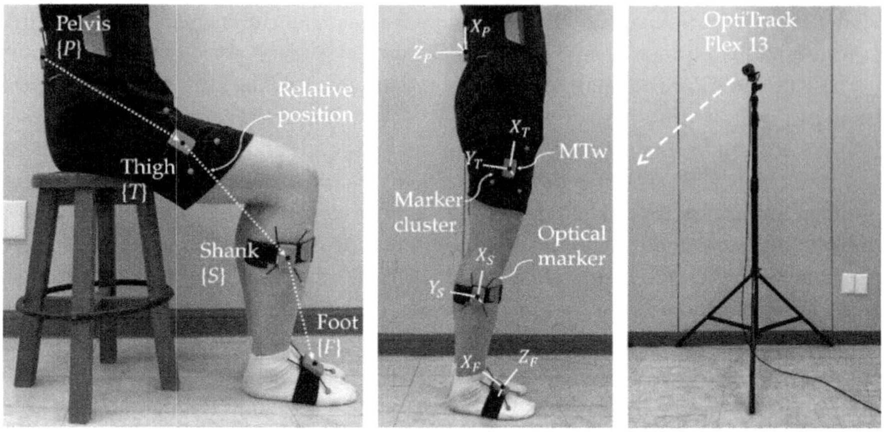

(a) A wearable posture estimation system [30]

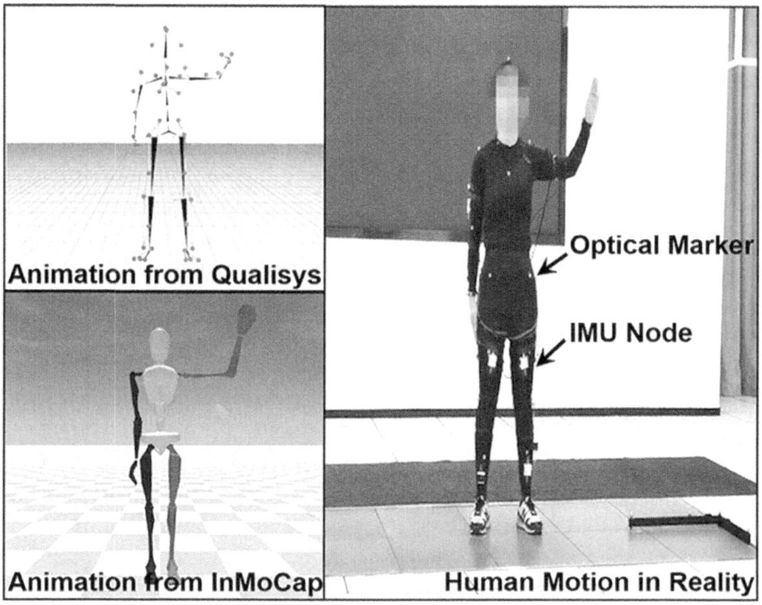

(b) Digital twin based on motion data [11]

Fig. 1.3 Attitude calculation methods of wearable systems. **a** A wearable posture estimation system [30], **b** Digital twin based on motion data [11]

through hub controllers, communication modules, and battery modules, ultimately achieving remote control of robots based on this digital twin model (Fig. 1.3).

The accuracy of the human posture solution is disturbed by external acceleration and measurement noise. However, modeling external acceleration and fine-tuning noise covariance can effectively reduce the error. Fan et al. initially employed a quaternion-based complementary filter as the primary attitude solution algorithm [34]. This approach is resistant to magnetic interference and includes two distinct tuning parameters designed to handle external acceleration and magnetic disturbances. They subsequently introduced an adaptive strategy based on finite state machines, developing two separate state machines to address external acceleration and magnetic field interference in attitude fusion. Ahmed et al. designed a Kalman filter to estimate the direction of linear acceleration, addressing the issue in traditional methods where significant information from sensor axes without linear acceleration is often disregarded [35]. By assigning weights to the acceleration in the filter and retaining the motion data of key axes, the accuracy of attitude calculation was effectively improved, although the algorithm paid insufficient attention to yaw angle and attitude. Shi et al. proposed a novel method for hybrid upper limb motion tracking and 3-D positioning by fusing the IMU and UWB systems. We first simplified the human upper limb as a kinematics model and employed the quaternion method to calculate the attitude angle of each segment. To compensate for the accumulated error of gyroscope measurement, the fourth-order Runge–Kutta Madgwick orientation filter was adopted to improve the accuracy of 3-D motion tracking through the optimal fusion of the accelerometer, gyroscope, and magnetometer of the MARG system (Fig. 1.4) [36]. Lee et al. expressed the external acceleration of the filter as a white noise process of first-order low-pass filtering, which was then integrated into the fusion algorithm [37]. However, the filter employed a constant cutoff constant, which limited its ability to obtain the optimal attitude estimation result in dynamic motion. McGrath et al. proposed a magnetometer-free estimation method, introducing a new model of a knee joint hinge [38]. It includes segment length anthropometry, segment cross-leg length discrepancy, and the relationship between the knee axis and femur/tibia segment. Then the maximum posteriori problem is formulated as a factor graph and inference is performed via post-hoc, on-manifold global optimization. Remarkably, despite the absence of a magnetometer, the accuracy of knee flexion angle estimation using this method is comparable to that of the most advanced methods that incorporate a magnetometer (Fig. 1.4).

(2) Human posture calibration

Human posture calibration is the premise of human posture reconstruction, which aims to establish the transformation relationship between the sensor coordinate system and limb segment coordinate system. Commonly used human posture calibration methods mainly include four categories: the manual alignment methods, the functional calibration methods, the model-based calibration methods, and the augmentation data methods [39].

(a) The manual alignment method is a technique used in IMU motion capture systems. It requires the IMU to be directly attached to the body's limb segments and ensures that the sensor coordinate system of the IMU precisely aligns with the body coordinate system of the corresponding limb segment. This method primarily relies on manual operation to achieve precise coordinate system

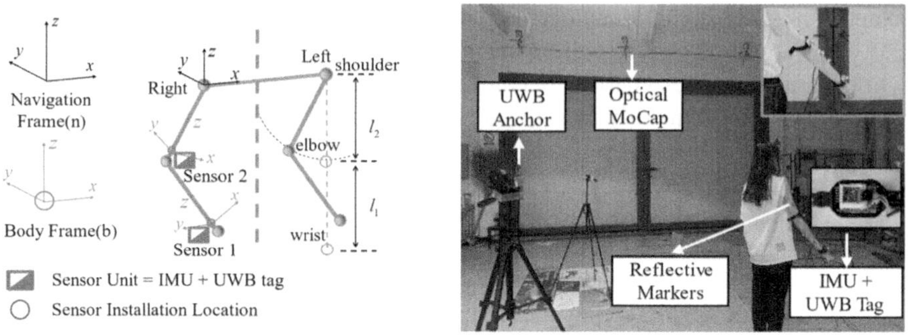

(a) Movement along single axis [36]

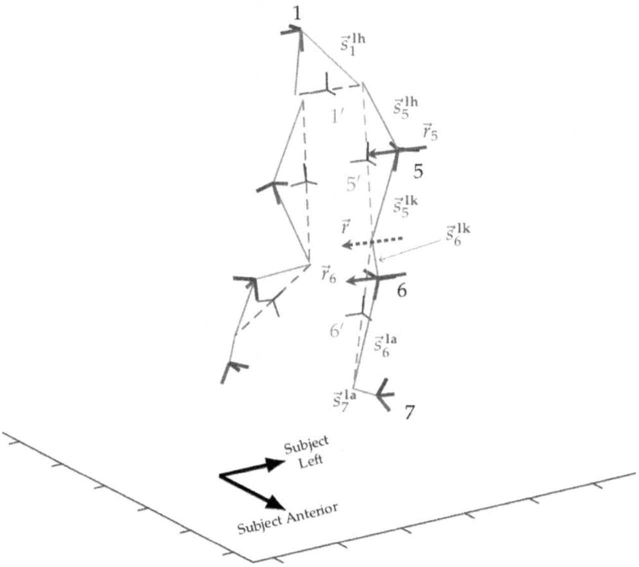

(b) The human-IMU kinematic system, with the subject mid-stride [38]

Fig. 1.4 Methods of improving the accuracy of the human motion solution. **a** Movement along single axis [36], **b** The human-IMU kinematic system, with the subject mid-stride [38]

matching. Specifically, the operator needs to ensure that the direction and position of each IMU device are consistent with the natural direction and anatomical structure of the limb segment being measured. Sun et al. adopted a simplified human model and then determined the transformation relationship between the sensor coordinate system and the human coordinate system by manual alignment method, which was used to calculate the joint angle of lower limbs [40] (Fig. 1.5). This kind of method requires high accuracy of sensor installation and is easily disturbed by dynamic motion. While the artificial alignment method is tedious to operate and requires significant technical capability from the operator, it provides a feasible solution for motion capture in the absence of advanced automatic calibration systems. This method is particularly suitable for occasions with budget constraints or special equipment requirements. Although its precision is not as high as automated systems, it remains effective for some basic motion capture tasks.

(b) The functional calibration method is a technique employed in inertial motion capture systems. It automatically determines the direction of the limb segment's coordinate axis by making the wearer perform a series of specific calibration actions. This method improves the accuracy of motion data utilizing the wearer's natural movement to identify and correct the sensor's placement position and direction. Fasel et al. used the wearer's upright static posture data to determine the upper-lower axial direction of the body coordinate system and applied the slow squat movement data to determine the inner-outer axial direction of the body coordinate system, then the remaining axial direction was determined by pointing cross product [41] (Fig. 1.5). Lebleu et al. asked the wearer to complete the designated posture of lower limbs calibration, and demonstrated the equivalence of the three calibration movements by experiments, as shown in Fig. 1.5 [42]. The advantage of the functional calibration method is its high degree of automation. It can quickly and accurately complete sensor calibration, reducing the need for manual operation. Moreover, since this method relies on the wearer's natural movement, it can better adapt to individual differences in actual applications, and then improves the personalized accuracy of motion capture. However, the soft tissue artifacts between the inertial sensor and the wearing part, as well as irregular calibration postures, will affect the accuracy of the functional calibration method.

(c) The model-based calibration methods play a crucial role in inertial motion capture systems. These methods use kinematic or statistical models of joints to estimate the direction of limb segment coordinate axes, thus achieving accurate interpretation and adjustment of sensor data. The advantage of model-based calibration methods lie in their high degree of automation and precision. As these methods rely on detailed models and algorithms, they can provide more accurate, more stable calibration results compared to traditional manual calibration methods. Moreover, model calibration methods are capable of accommodating individual differences, providing personalized calibration services for each user. Zimmermann et al. used the deep learning model to determine the transformation relationship between the coordinate system of the human lower limb sensor and the coordinate system of the limb segment [43] (Fig. 1.5). However, training the deep learning model necessitates a large dataset of inertial motion samples.

(d) The augmentation data method provides more accurate and comprehensive motion capture data by integrating devices other than IMUs. The crux of this method lies in using additional sensors or systems to determine the transformation relationship between the sensor coordinate system and the limb segment coordinate system, thereby improving the accuracy and reliability of motion

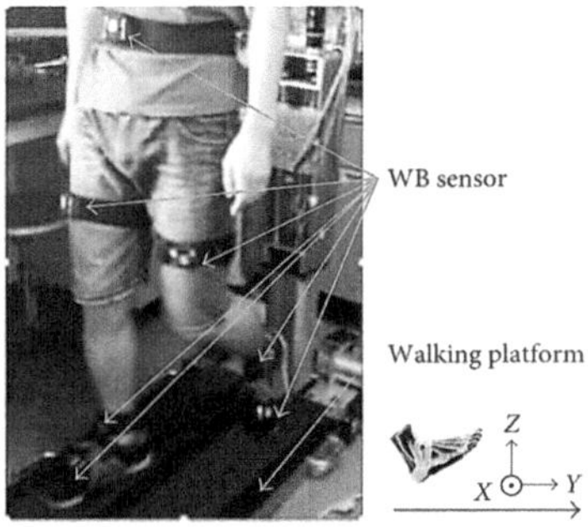

(a) Manual alignment method [40]

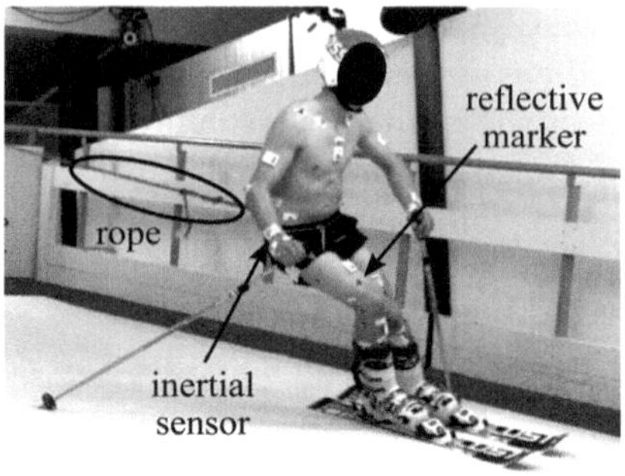

(b) Functional calibration method for skiing [41]

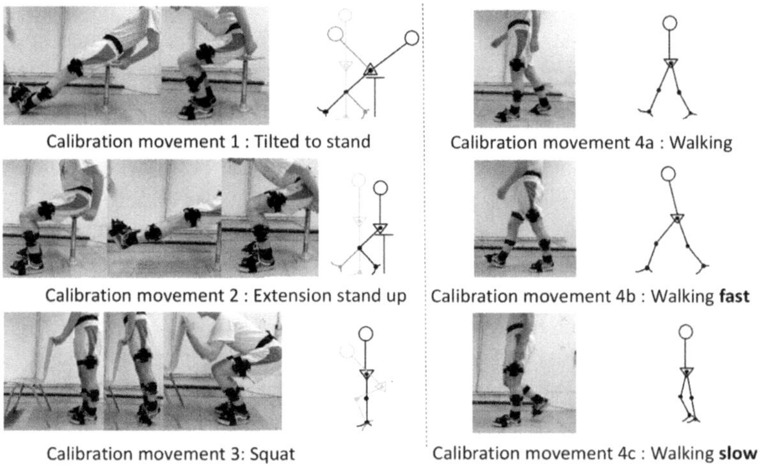

(c) Calibration action based on functional calibration method [42]

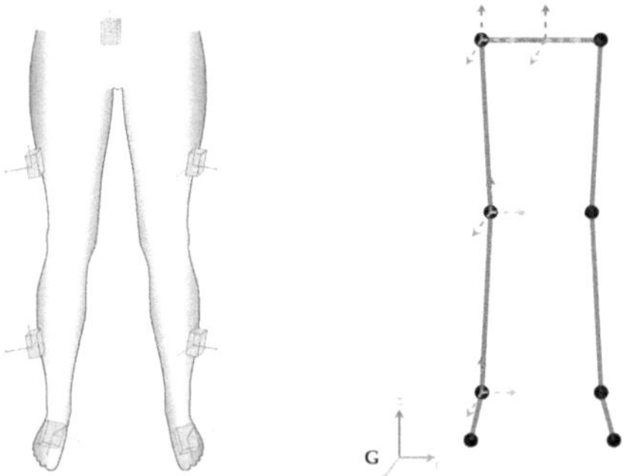

(d) Sensor coordinate system and limb segment coordinate system of human lower limbs [43]

Fig. 1.5 Human posture calibration methods, **a** Manual alignment method [40], **b** Functional calibration method for skiing [41], **c** Calibration action based on functional calibration method [42], **d** Sensor coordinate system and limb segment coordinate system of human lower limbs [43]

capture. Alizadegan et al. introduced unmarked motion capture equipment to measure the positions of elbows and wrists relative to shoulders in the global coordinate system, and then artificially corrected the error in human posture calibration [44]. However, the use of additional calibration equipment reduces the convenience of attitude calibration.

In response to the demand of human motion perception and analysis, accurate external acceleration modeling and noise estimation are paramount for improving the accuracy of human posture calculation. What's more, developing convenient and effective calibration methods is the focus of human posture reconstruction. In conclusion, further research is warranted to refine the perception and analysis of human motion based on inertial motion data.

1.2.2 Research on Human Interactive Behavior

(1) Human motion intention capture

Human motion intent capture is a technique for anticipating subsequent actions by monitoring and analyzing human movement. It typically employs sensors, computer vision, artificial intelligence, etc., to identify and interpret patterns and signals of human motion. By conveying the intention of human movement to robots, we can effectively leverage the respective advantages of humans and machines to accomplish the goal and task together.

Sinha et al. used two connecting rods and a potentiometer which was fixed at the joint to form a simple exoskeleton device [45]. The limb movement drove the upper and lower connecting rods to move alternately, while the joint data were determined by the output of the potentiometer, and then it was mapped to the joint servo motor at the robot end by the microcontroller. Fan et al. used a CNN-based transfer learning strategy was proposed to improve the recognition accuracy of amputees, whose signals are weaker, and that have higher noise and are less available in the databases (Fig. 1.6) [46]. Li et al. used a 6 DOFs tactile device to control the position of the manipulator and to toggle the end gripper on the Baxter robot [47]. They also utilized two computers in tandem to facilitate bidirectional remote visualization, thereby markedly enhancing the user's operational experience. Schwarz built a teleoperation control system based on an immersive 3D visualization system with force feedback equipment, which used the robot itself at the control end as the human motion capture device to collect the motion data of arms and fingers, then teleoperated the robot [48]. This system has obvious advantages in completing complex objectives and tasks. Tran et al. proposed a control method for a CDHR robot in a teleoperation system was proposed using the HIL simulation technique. The teleoperation system consists of a haptic device, called a delta robot, and a CDHR robot, and an HIL simulation system was developed to evaluate the safety of the teleoperation control algorithm and system before implementing it at the experimental scale (Fig. 1.6) [49]. The traditional method can accurately obtain the human operation intention and is suitable for complex application scenarios, but it tends to be costly and may not align with natural human operational habits. Teleoperation that utilizes human body language aligns with human operation habits and has great application potential. Stanton et al. used Xsens MVN motion capture equipment to collect the motion data of main limb segments [50]. They used a feedforward neural network and particle swarm optimization algorithm to determine the mapping relationship between motion data and robot joint motion, which realized human–robot teleoperating interaction. This method does not need kinematics modeling but needs to build a human–robot motion mapping dataset. Zhu et al. presented a method of IMU motion capture with adaptive filtering in the teleoperation system. The method captures the motion of the human upper limbs and provides the position and direction information of the end of the arm, which can be mapped onto the end of the robot after adaptive filtering to complete the control (Fig. 1.6) [51]. Zhu et al. integrated an inertial motion capture system with the surface electromyogram (EMG) signal to collect data from the armband, and used an IMU to obtain the pose information of the arm end [31]. Then they fused the surface EMG signal to reduce the motion error caused by the physiological tremor of the human body, which indirectly improved the tracking performance of the robot. Li et al. proposed a human intent prediction method based on a dual-stream structure using RGB images and optical flow [52]. This method calculates designated parameters to determine the starting frame of each attempt at human action. What's more, it uses RGB images and optical flow to train spatial and temporal networks to predict human intent. The prediction results from the spatial and temporal networks are then fused using an averaging fusion method. Dafarra et al. set up a complex data-collecting system, which can obtain information such as human movements, sounds, expressions, and tactile sensations [53]. With the help of this system, the operator can teleoperate with the humanoid robot iCub and interact with others through this robot. With the improvement of system integration complexity, the monitoring of human motion intention is becoming more and more complete.

(2) Mapping of human motion

With the development of robot technology and artificial intelligence, research into human–robot interaction occupies an increasingly important position in the field of modern technology. In teleoperation systems, accurately mapping human motion to robots or other devices is a key area of research interest. The human motion mapping strategy is one of the research focuses on human–robot teleoperation. Considering the kinematics difference between the human arm and the robot arm, Shi et al. proposed a variable coefficient method based on a weighted augmented Jacobian matrix, which made the slave robot with one arm and the master robot maintain similar motion configuration by simulating joint motion [54]. Fani et al. put forward a method to enhance teleoperation, which combines wearable tactile feedback equipment with remote impedance technology [55]. Through the integrated interface, the operator can adjust the stiffness and posture of the manipulator system through interactive force and visual perception. The above researches primarily use one-arm robots with limited natural motion, which is not convenient for performing complex interactive

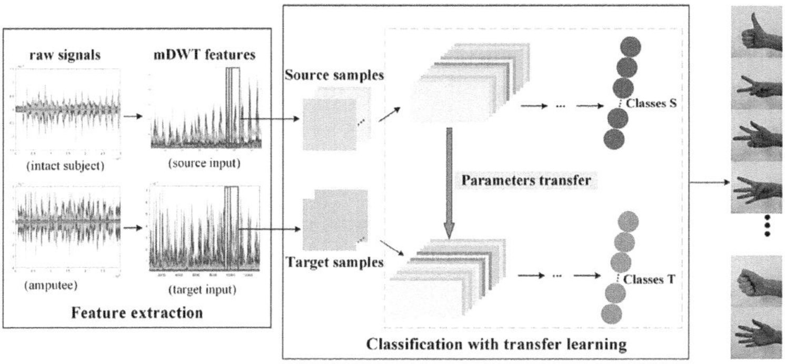

(a) A workflow of the motion intention recognition system [46]

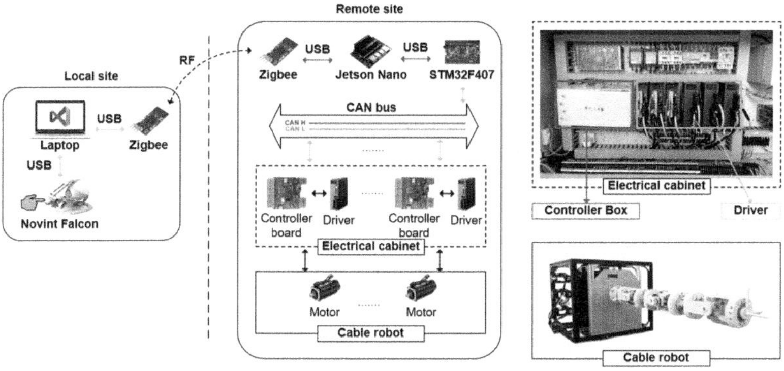

(b) Teleoperate mapping between Novint Falcon and CDHR robot [49]

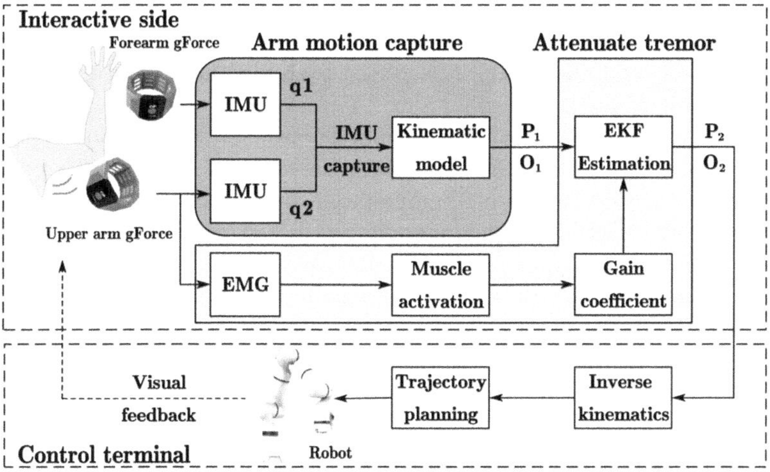

(c) Using principle of the teleoperated robot system [51]

Fig. 1.6 Human motion intention capture systems. **a** A workflow of the motion intention recognition system [46], **b** teleoperate mapping between Novint Falcon and CDHR robot [49], **c** using principle of the teleoperated robot system [51]

tasks. Dual-arm teleoperation represents an emerging trend in human–robot interaction. Kruse et al. integrated a control system of a dual-arm industrial robot, which used the mixed position/force control method to obtain the interactive force information [56]. It enabled the robot system to autonomously identify and grab target objects, even detecting the operator's operation gestures through Kinect to determine different operation modes.

The reason why planning human-like motions for robot arms becomes a difficult point in motion mapping strategies is due to its inherent complexity and diversity. It must precisely simulate the multiple joints and high degrees of freedom of human hands, to achieve their wide range of motion. Furthermore, the human hand can perform extremely delicate operations, and the robot arm must maintain high precision and stability when simulating these operations, which is very challenging in terms of technical implementation. The robot arm must also address issues of real-time responsiveness and adaptability, necessitating swift reactions to environmental changes and real-time movement adjustments. This requires its control algorithms to be efficient and robust. In addition, hand motions often need to coordinate with other body parts, which requires that the motion planning of the robot arm integrates seamlessly with other body movements. At the same time, safety and robustness are critical considerations to ensure that the robot arm operates without causing harm to humans or the environment and maintains stability amidst uncertainties. The current sensing and control technologies may not yet fully meet these demands, limiting the performance of robot arms. Establishing a mathematical model that can accurately simulate the biological characteristics of the human hand and developing corresponding efficient algorithms remains a significant challenge. This is because these algorithms need to fully consider the dynamic characteristics of the robot arm, environmental constraints, and task objectives. Finally, to ensure users can easily understand and accept the operation mode of the robot arm, the motion generated by the system must conform to human intuition and habits, which is also an indispensable aspect of the motion planning of humanoid robot arms. Ayusawa et al. put forward a relocation method to solve the problems of human motion reproduction and robot motion planning [57]. They obtained the joint trajectory of the human body under the redirection constraint and mapped the captured motion to the humanoid robot so that the robot could imitate and track the human motion. Škulj et al. presented a novel wireless wearable system that uses only IMUs to determine the orientation of the operator's upper body parts. An algorithm was developed to transform the measured orientations to movement commands for an industrial collaborative robot. The algorithm includes a calibration procedure, which aligns the coordinate systems of all IMUs, the operator, and the robot, and the transformation of the operator's relative hand motions to the movement of the robot's end effector, which takes into account the operator's orientation relative to the robot. The developed system is demonstrated with an example of an industrial application in which a workpiece needs to be inserted into a fixture. The robot's motion is compared between the developed system and a standard robot controller. The results confirm that the developed system is intuitive, allows for flexible control, and is robust enough for use in industrial collaborative robotic applications (Fig. 1.7) [58]. García et al. used the potential field method to represent the human motion trajectories and designed a planning strategy which was based on the progressive optimal fast random search tree algorithm, to minimize the path cost function and map the human motion into a dual-arm robot motion [59]. Wang et al. proposed a method based on virtual joints, using the human arm movement data obtained from an inertial motion capture system [60]. This facilitates a quick and intuitive kinematic mapping between the human arm and the UR5 robot, thereby leading to a robotic imitation system that can simultaneously mimic the end position and configuration of the human arm. In addition, some researchers designed motion mapping strategies by analyzing the structure and motion characteristics of human arms. Xie et al. constructed hypotheses explaining the natural motion characteristics of human arms [61]. They also developed inverse kinematics algorithms of robots and realized humanoid motion planning. Dìng et al. constructed a triangular model of the human arm and derived the forward and inverse kinematics relationship among joint space, arm triangular space, and task space, obtaining the motion trajectory of arm joint [62]. Wang et al. decomposed the arm motion into a set of different motion processes [63]. They designed motion models and attitude prediction indexes at different planning levels according to the motion characteristics of each planning level and finally realized the humanoid motion planning of the robot arms.

(3) Human motion estimation

Currently, many operational tasks rely on manual labor, which can be labor-intensive and technically challenging. Human–robot interaction provides a feasible solution, combining the flexibility of humans with the efficiency of machines, significantly influencing work efficiency, resource coordination, and task distribution. Through human–robot coordination, work efficiency can be effectively improved. Robots can continuously perform repetitive or physically labor-intensive tasks with higher speed and precision than humans, thus notably enhancing overall efficiency. In terms of resource coordination, human–robot interaction optimizes the allocation and scheduling of human resources. The introduction of robots offers more flexibility and possibilities for resource coordination, enabling dynamic adjustments to robots' work content and intensity, and realizing optimal resource allocation. Regarding task distribution, human–robot collaboration leverages the computational and data analysis capabilities of robots, allowing for more precise and reasonable distribution, enhancing the quality and efficiency of task completion, and reducing error and delay rates. Human motion estimation is a pivotal step in coordinating human–robot cooperation. According to the obtained human motion intention, it triggers the robot to determine "when and where to send the target object" and performs operational tasks to realize interaction [64].

Motion estimation methods based on wearable sensors mainly include methods based on machine learning models or deep learning models. Xuan et al. equipped the human stump with accelerometers, gyroscopes, and pressure sensors, then they combined with correlation coefficient analysis, data fusion algorithm, and hidden Markov model to predict human motion intentions [65]. Wang used sensing data from IMUs and angle sensors to classify human motion intentions by using support

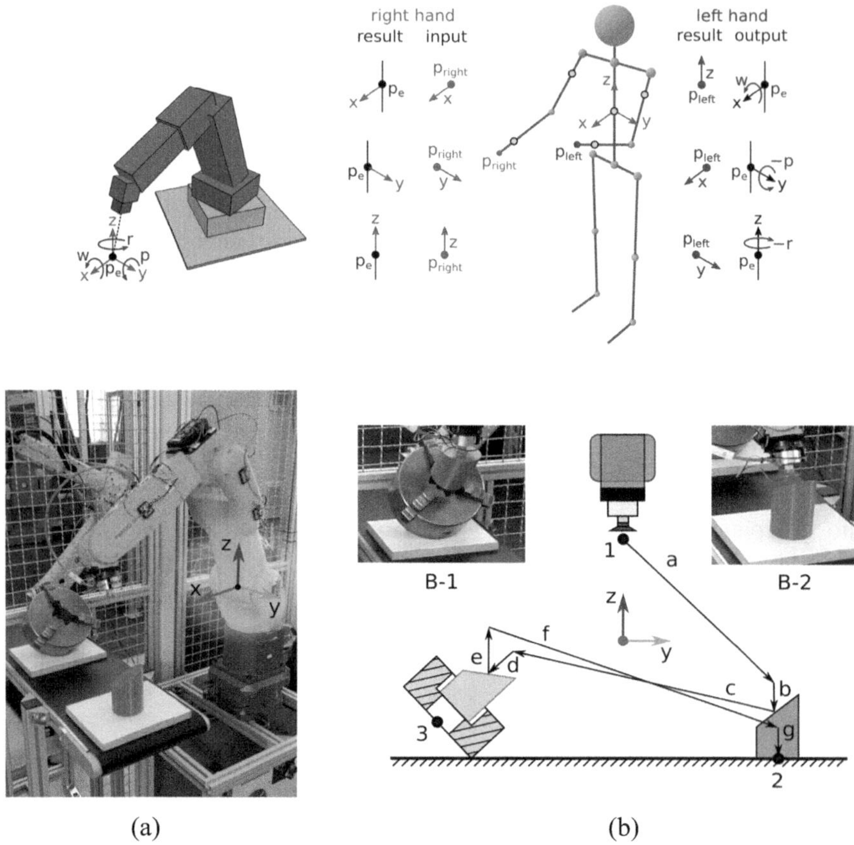

Fig. 1.7 Redirected human motion to robot [58]

vector machine model [66]. With the development of artificial intelligence algorithms, researchers introduced deep learning models into the application of human posture estimation based on inertial motion data. IONet conceptualized the tracking problem as a physical model based on sequence and estimated the motion trajectory by using a bidirectional long-term and short-term memory model [67]. In addition to the application of gait trajectory tracking, Yang et al. introduced the deep learning algorithm into arm trajectory tracking [68]. They built a motion-tracking framework based on the Elman neural network and obtained attitude information by using acceleration and angular velocity. They put preprocessed data into the network and then got predicted position information (Fig. 1.8). Wei et al. proposed a motion tracking model based on the motion data of a single IMU, utilizing a recurrent neural network as the foundational framework and incorporating the initial joint position and velocity as the model's initial state [69]. Wei et al. employed a Recursive Neural Network (RNN) model to estimate the 3D positions of the wrist and elbow from noisy IMU data [70]. This method allowed them to track the user's 3D arm movements in real-time

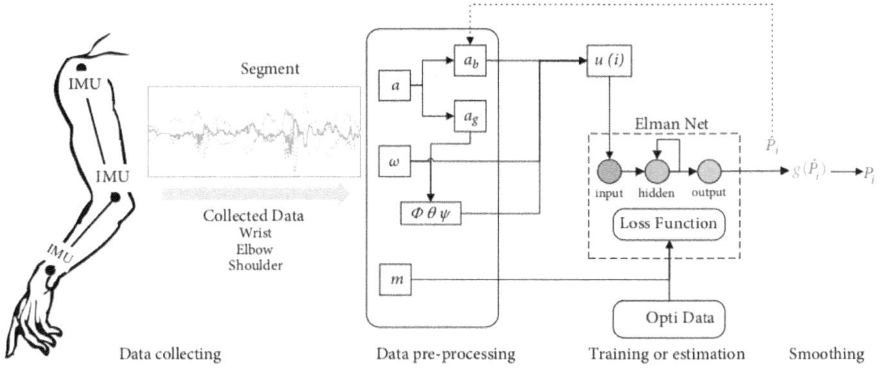

Fig. 1.8 Using acceleration and angular velocity to predict position information [68]

using a single 6-axis IMU sensor on a consumer-grade smartwatch, without the need for any magnetometers. To improve the universality of the network model, Lin et al. combined residual networks with bidirectional long-term and short-term memory networks, in order to track three-dimensional motion trajectories [71]. The model takes the inertial motion data from a single IMU as input, then estimates and outputs three-dimensional position data, which performs well on both hand motion datasets and gait walking datasets, so it can be used to estimate the trajectory of human daily activities. The application of deep learning algorithms is anticipated to compensate for the limited environmental adaptability of IMUs, the robustness shortcomings of traditional algorithm models, and their constrained application scenarios.

Addressing the complexity of the usage scenes, the high dimension of joint motion and the uncertainty of interactive behaviors are significant challenges within the domain of human motion perception and estimation. Traditional human motion estimation algorithms have some shortcomings, such as poor robustness of motion modeling and tedious manual adjustment of parameters. Incorporating deep learning algorithms can effectively alleviate the above problems, but the research on human motion estimation based on deep learning models requires further exploration.

1.3 Chapter Overview

This book focuses on key technologies such as human motion perception calculation based on IMUs, intuitive human–robot teleoperation based on IMU-based motion perception, and enhanced human–robot collaboration through motion estimation. Specifically, the concept of digital twins is extended from the physical system to human beings, focusing on human movement. Based on the construction of a human motion digital twin, we perceive the inertial motion data of the human body and quantify and analyze the key motion parameters. What's more, we also analyze the characteristics and attributes of human motion and represent the human interaction behavior in combination with human–robot interaction application scenarios. Finally, on the basis of the self-developed inertial motion capture system and the human–robot interaction system, the above algorithm framework is evaluated and verified. The specific structural arrangement is as follows.

This chapter: Introduction. Firstly, the research background and significance of digital twin technology in relation to human movement in the smart manufacturing are explored. The advancement of smart manufacturing has heightened the focus on enhancing the efficiency and accuracy of human–robot collaboration. The application of digital twin technology to human motion research offers substantial support for improving human–robot interaction in this context. Secondly, the book reviews the current status of key technologies essential for developing human motion digital twin systems. It also addresses the challenges that need to be resolved in this area. Finally, the main research content and chapter organization of this book are outlined. It emphasizes the in-depth exploration of human motion digital twin technology within the context of smart manufacturing. This provides both theoretical insights and practical guidance aimed at advancing human–robot collaboration in smart manufacturing.

Chapter 2: IMU-based human motion perception. This chapter explores key technologies in human motion perception and computation, with a focus on pose estimation and posture calibration. An adaptive extended Kalman filter is proposed for pose estimation, using quaternions to represent orientations and transforming three-dimensional rotations into an optimization problem solved via nonlinear least squares. The algorithm incorporates an adaptive strategy that adjusts the process noise covariance matrix in real time. It improves accuracy by compensating for sensor noise and external disturbances. In addition, a novel attitude calibration algorithm based on kinematic constraints is introduced for human motion reconstruction. This algorithm uses three specific postures to establish the relationship between the sensor and body part coordinate systems. By incorporating kinematic constraints, it reduces calibration errors, improving motion capture accuracy and reliability in smart manufacturing environments. The effectiveness of the proposed algorithms is validated through integration into a self-developed inertial motion capture system. Experimental results show significant improvements in pose estimation accuracy, stability, and overall motion reconstruction. These findings support the development of high-precision, cost-effective human motion capture systems for smart manufacturing.

Chapter 3: Intuitive human–robot teleoperation based on IMU-based motion perception. This chapter researches human motion mapping technology for teleoperation in smart manufacturing. We track joint poses and calculate their motion trajectories, ensuring accuracy and real-time performance through advanced data processing algorithms using inertial motion capture. Then we developed human and robot skeletal models, calculated their motion range overlap, and designed a master–slave pose mapping strategy for natural robot movement during teleoperation. In a practical application, we implemented a teleoperation system in a laboratory setting, conducting remote experiments between China and Sweden. These experiments validated the proposed models and strategies, demonstrating the adaptability and stability of the teleoperation technology in different environments. The results highlight the effectiveness of the human motion mapping technology for developing human motion digital twin systems and provide new approaches for human motion analysis, human–robot interaction, and automated production in smart manufacturing.

Chapter 4: Enhanced human–robot collaboration through motion estimation. This chapter outlines the process of motion estimation and reconstructing human postures. High-precision inertial sensors capture 3D movement data for perception and analyzing human movements. In the pre-processing phase, advanced algorithms eliminate sensor errors and external interferences, ensuring reliable data. A conversion algorithm then transforms this data into human posture information, providing a reference for acquiring robot instructions. Time series analysis is used to predict future movements, enhancing real-time robot adaptability to human actions in tasks like precision assembly. Integrating these algorithms into an inertial motion capture system allows real-time capture and conversion of human actions into robot instructions, improving collaboration efficiency in dynamic production environments.

Chapter 5: Conclusion and outlook. This chapter summarizes the key research presented in the book, focusing on human motion capture and its application in smart manufacturing. The research explores techniques for accurate motion data acquisition, pose estimation, motion trajectory prediction, and the integration of these methods into teleoperation systems to improve human–robot collaboration. These advancements contribute to optimizing manufacturing processes, enhancing productivity, and ensuring worker safety. Looking ahead, future research should focus on improving the accuracy, robustness, and cost-effectiveness of motion capture systems, particularly in real-world conditions. Further advancements in data processing algorithms, incorporating artificial intelligence, will enable better real-time adaptation of robots to human actions, enhancing collaboration. Integrating multi-sensory data fusion and more detailed biomechanical models could improve motion capture accuracy and provide deeper insights into human ergonomics and performance. Moreover, expanding the scalability and interoperability of these systems across different industries is crucial. Developing standardized frameworks for human motion capture and teleoperation will facilitate widespread adoption and integration in diverse manufacturing environments.

References

1. Xu X, Lu Y, Vogel-Heuser B, Wang L (2021) Industry 4.0 and Industry 5.0–Inception, conception, and perception. J Manuf Syst 61:530–535
2. Zhou J, Zhou Y, Wang B, Zang J (2019) Human-cyber-physical systems (HCPSs) in the context of new-generation smart manufacturing. Engineering 5(4):624–636
3. Wang B, Zhou H, Li X et al (2024) Human digital twin in the context of Industry 5.0. Robot Comput-Integr Manuf 85:102626
4. Lei T, Rong Y, Wang H, Huang Y, Li M (2020) A review of vision-aided robotic welding. Comput Ind 123:103326
5. Zhang Q, Xiao R, Liu Z, Duan J, Qin J (2023) Process simulation and optimization of arc welding robot workstation based on Digital Twin. Machines 11(1):53

6. Chen S, Liu J, Chen B, Suo X (2022) Universal fillet weld joint recognition and positioning for robot welding using structured light. Robot Comput-Integr Manuf 74:102279
7. Zhang Z, Ren W, Yang Z, Wen G (2020) Real-time seam defect identification for Al alloys in robotic arc welding using optical spectroscopy and integrating learning. Measurement 156:107546
8. Tsuzuki R (2021) Development of automation and artificial intelligence technology for welding and inspection process in aircraft industry. Weld World 66:105–116
9. Wang B, Hu SJ, Sun L, Freiheit T (2020) Intelligent welding system technologies: state-of-the-art review and perspectives. J Manuf Syst 56:373–391
10. Tao F, Zhang H, Liu A, Nee AYC (2019) Digital twin in industry: state-of-the-art. IEEE Trans Industr Inf 15(4):2405–2415
11. Zhou H, Lv H, Wang R, Wu H, Yang G (2023) Revitalizing human-robot interaction: phygital twin driven robot avatar for China-Sweden teleoperation. Chin J Mech Eng 36(1):124
12. Rhodin H, Richardt C, Casas D et al (2016) EgoCap: egocentric marker-less motion capture with two fisheye cameras. ACM Trans Graph 35(6):1–11
13. Fu D, Kuduvalli G (2006) Enhancing skeletal features in digitally reconstructed radiographs. In: Image processing, San Diego, CA, p 61442M
14. Pfister A, West AM, Bronner S, Noah JA (2014) Comparative abilities of Microsoft Kinect and Vicon 3D motion capture for gait analysis. J Med Eng Technol 38(5):274–280
15. Gutierrez M, Gomez B, Retamal G et al (2024) Comparing optical and custom IoT inertial motion capture systems for manual material handling risk assessment using the NIOSH lifting index. Technologies 12(10):180
16. Baskwill AJ, Belli P, Kelleher L (2017) Evaluation of a gait assessment module using 3d motion capture technology. Int J Ther Massage Bodywork Res Educ Pract 10(1):3–9
17. Espitia-Mora LA, Vélez-Guerrero MA, Callejas-Cuervo M (2024) Development of a low-cost markerless optical motion capture system for gait analysis and anthropometric parameter quantification. Sensors 24(11):3371
18. Napoli A, Glass S, Ward C, Tucker C, Obeid I (2017) Performance analysis of a generalized motion capture system using Microsoft Kinect 2.0. Biomed Signal Process Control 38:265–280
19. Brito AV, Fonseca P, Costa MJ et al (2024) The influence of kinematics on tennis serve speed: an in-depth analysis using Xsens MVN biomech link technology. Bioengineering 11(10):971
20. Roetenberg D, Luinge H, Slycke P (2009) Xsens MVN: full 6DOF human motion tracking using miniature inertial sensors. Xsens Motion Technologies BV, Tech Rep 1
21. Raghu SL, Kang C, Whitehead P, Takeyama A, Conners R (2019) Static accuracy analysis of Vicon T40s motion capture cameras arranged externally for motion capture in constrained aquatic environments. J Biomech 89:139–142
22. Yang C, Jiang M, Chen M, Fu M, Li J, Huang Q (2021) Automatic 3-D imaging and measurement of human spines with a robotic ultrasound system. IEEE Trans Instrum Meas 70:1–13
23. Zhou H, Yang G, Lv H, Huang X, Yang H, Pang Z (2020) IoT-enabled dual-arm motion capture and mapping for telerobotics in home care. IEEE J Biomed Health Inform 24(6):1541–1549
24. Farahan SB, Machado JJM, De Almeida FG, Tavares JMRS (2022) 9-DOF IMU-based attitude and heading estimation using an extended Kalman filter with bias consideration. Sensors 22(9):3416
25. Candan B, Soken HE (2021) Robust attitude estimation using IMU-only measurements. IEEE Trans Instrum Meas 70:1–9
26. Wu J, Zhou Z, Chen J, Fourati H, Li R (2016) Fast complementary filter for attitude estimation using low-cost MARG sensors. IEEE Sens J 16(18):6997–7007
27. Majumder S, Deen MJ (2021) Wearable IMU-based system for real-time monitoring of lower-limb joints. IEEE Sens J 21(6):8267–8275
28. Yuan Q, Asadi E, Lu Q, Yang G, Chen I-M (2019) Uncertainty-based IMU orientation tracking algorithm for dynamic motions. IEEE/ASME Trans Mechatron 24(2):872–882
29. Lisini Baldi T, Farina F, Garulli A, Giannitrapani A, Prattichizzo D (2020) Upper body pose estimation using wearable inertial sensors and multiplicative Kalman filter. IEEE Sensors J 20(1):492–500

30. Lee CJ, Lee JK (2022) Wearable IMMU-based relative position estimation between body segments via time-varying segment-to-joint vectors. Sensors 22(6):2149
31. Zhu Z, Sorkhabadi SMR, Gu Y, Zhang W (2022) Invariant extended Kalman filtering for human motion estimation with imperfect sensor placement. In: American control conference (ACC), pp 3012–3018
32. Zhang R, Yang H, Höflinger F, Reindl LM (2017) Adaptive zero velocity update based on velocity classification for pedestrian tracking. IEEE Sens J 17(7):2137–2145
33. Zandbergen MA, Reenalda J, Van Middelaar RP, Ferla RI, Buurke JH, Veltink PH (2022) Drift-free 3D orientation and displacement estimation for quasi-cyclical movements using one inertial measurement unit: application to running. Sensors 22(3):956
34. Fan B, Li Q, Liu T (2018) Improving the accuracy of wearable sensor orientation using a two-step complementary filter with state machine-based adaptive strategy. Meas Sci Technol 29(11):115104
35. Ahmed H, Tahir M (2017) Improving the accuracy of human body orientation estimation with wearable IMU sensors. IEEE Trans Instrum Meas 66(3):535–542
36. Shi Y, Zhang Y, Li Z, Yuan S, Zhu S (2023) IMU/UWB fusion method using a complementary filter and a Kalman filter for hybrid upper limb motion estimation. Sensors 23(15):6700
37. Lee JK, Park EJ, Robinovitch SN (2012) Estimation of attitude and external acceleration using inertial sensor measurement during various dynamic conditions. IEEE Trans Instrum Meas 61(8):2262–2273
38. McGrath T, Stirling L (2020) Body-worn IMU human skeletal pose estimation using a factor graph-based optimization framework. Sensors 20(23):6887
39. Vitali RV, Perkins NC (2020) Determining anatomical frames via inertial motion capture: a survey of methods. J Biomech 106:109832
40. Sun T, Li H, Liu Q et al (2017) Inertial sensor-based motion analysis of lower limbs for rehabilitation treatments. J Healthcare Eng 2017:1–11
41. Fasel B, Spörri J, Schütz P, Lorenzetti S, Aminian K (2017) An inertial sensor-based method for estimating the athlete's relative joint center positions and center of mass kinematics in alpine ski racing. Front Physiol 8:850
42. Lebleu J, Gosseye T, Detrembleur C, Mahaudens P, Cartiaux O, Penta M (2020) Lower limb kinematics using inertial sensors during locomotion: accuracy and reproducibility of joint angle calculations with different sensor-to-segment calibrations. Sensors 20(3):715
43. Zimmermann T, Taetz B, Bleser G (2018) IMU-to-segment assignment and orientation alignment for the lower body using deep learning. Sensors 18(1):302
44. Alizadegan A, Behzadipour S (2017) Shoulder and elbow joint angle estimation for upper limb rehabilitation tasks using low-cost inertial and optical sensors. J Mech Med Biol 17(02):1750031
45. Sinha AK, Sahu SK, Bijarniya RK, Patra K (2017) An effective and affordable technique for human motion capturing and teleoperation of a humanoid robot using an exoskeleton. In: 2nd international conference on man and machine interfacing (MAMI). IEEE, Bhubaneswar, pp 1–6
46. Fan J, Jiang M, Lin C et al (2021) Improving motion intention recognition for trans-radial amputees based on sEMG and transfer learning. Neural Comput Appl 35:16101–16111
47. Li J, Li Z, Hauser K (2017) A study of bidirectionally telepresent tele-action during robot-mediated handover. In: IEEE international conference on robotics and automation (ICRA), pp 2890–2896
48. Schwarz M, Lenz C, Rochow A, Schreiber M, Behnke S (2021) NimbRo Avatar: interactive immersive telepresence with force-feedback telemanipulation. In: IEEE/RSJ international conference on intelligent robots and systems (IROS), pp 5312–5319
49. Tran DT, Nguyen TD, Tran MK, Ahn KK (2024) Development of a hardware-in-the-loop platform for a teleoperation flexibility robotic system. Appl Sci 2024(5):2207
50. Stanton C, Bogdanovych A, Ratanasena E (2012) Teleoperation of a humanoid robot using full body motion capture, example movements, and machine learning. In: Australasian robotics and automation association. pp 1–10

51. Zhu H, Li X, Wang L et al (2022) IMU motion capture method with adaptive tremor attenuation in teleoperation robot system. Sensors 22(9):3353
52. Li S, Zhang L, Diao X (2020) Deep-learning-based human intention prediction using RGB images and optical flow. J Intell Rob Syst 97(1):95–107
53. Dafarra S, Pattacini U, Romualdi G et al (2024) iCub3 avatar system: enabling remote fully immersive embodiment of humanoid robots. Sci Robot 9(86):eadh3834
54. Shi Z, Huang X, Hu T, Tan Q, Hou Y (2016) Weighted augmented Jacobian matrix with a variable coefficient method for kinematics mapping of space teleoperation based on human-robot motion similarity. Adv Space Res 58(7):1401–1416
55. Fani S, Ciotti S, Catalano MG et al (2018) Simplifying telerobotics: wearability and teleimpedance improves human-robot interactions in teleoperation. IEEE Robot Autom Mag 25(1):77–88
56. Kruse D, Wen JT, Radke RJ (2015) A sensor-based dual-arm tele-robotic system. IEEE Trans Autom Sci Eng 12(1):4–18
57. Ayusawa K, Morisawa M, Yoshida E (2015) Motion retargeting for humanoid robots based on identification to preserve and reproduce human motion features. In: IEEE/RSJ international conference on intelligent robots and systems (IROS). pp 2774–2779
58. Škulj G, Vrabič R, Podržaj P (2021) A wearable IMU system for flexible teleoperation of a collaborative industrial robot. Sensors 21(17):5871
59. García N, Rosell J, Suárez R (2019) Motion planning by demonstration with human-likeness evaluation for dual-arm robots. IEEE Trans Syst Man Cybern Syst 49(11):2298–2307
60. Wang Z, Liang R, Chen Z, Liang B (2020) Fast and intuitive kinematics mapping for human-robot motion imitating: a virtual-joint-based approach. IFAC-PapersOnLine 53(2):10011–10018
61. Xie B, Zhao J, Liu Y (2011) Human-like motion planning for robot arm system. In: 15th international conference on advanced robotics (ICAR). pp 88–93
62. Ding X, Fang C (2013) A novel method of motion planning for an anthropomorphic arm based on movement primitives. IEEE/ASME Trans Mechatron 18(2):624–636
63. Wang C, Xia E, Zhao J, Xiong C, Liu J, Wei Y (2018) On human-like motion planning algorithm of anthropomorphic mechanical arms based on hierarchical planning strategy. J Graph 39(3):553–561
64. Zhou H et al (2023) An attention-based deep learning approach for inertial motion recognition and estimation in human-robot collaboration. J Manuf Syst 67:97–110
65. Xuan B, Liu Z, Chen L, Yang P (2017) Motion intention recognition and control of above knee prosthesis. J Southeast University (Nat Sci Ed) 47(6):1107–1116
66. Wang X (2019) Research on motion intention recognition technology for lower limb rehabilitation robot. Zhejiang University, Hangzhou, Zhejiang, China
67. Chen C, Lu X, Markham A, Trigoni N (2018) IONet: learning to cure the curse of drift in inertial odometry. Proc AAAI Conf Artif Intell 32(1):1
68. Yang H, Wang Y, Wang H et al (2022) Multi-inertial sensor-based arm 3D Motion tracking using Elman neural network. J Sens 2022:1–11
69. Wei W, Kurita K, Kuang J, Gao A (2021) Real-time limb motion tracking with a single IMU sensor for physical therapy exercises. In: 43rd annual international conference of the IEEE engineering in medicine and biology society (EMBC). pp 7152–7157
70. Wei W, Kurita K, Kuang J, Gao A (2021) Real-time 3D arm motion tracking using the 6-axis IMU sensor of a smartwatch. In: IEEE 17th international conference on wearable and implantable body sensor networks (BSN). IEEE, Athens, Greece, pp 1–4
71. Lin B-S, Lee I-J, Wang S-P, Chen J-L, Lin B-S (2022) Residual neural network and long short-term memory-based algorithm for estimating the motion trajectory of inertial measurement units. IEEE Sens J 22(7):6910–6919

Chapter 2
IMU-Based Human Motion Perception

Abstract Following the rule of human-centricity, human motion perception is the basis for constructing human motion digital twin. In particular, accurate motion pose calculation and pose calibration are the keys to reconstruct human motion. When capturing human data based on inertial motion capture system, it is still challenging to model motion and minimize noise. At the same time, the development of convenient and effective posture calibration methods is a prerequisite for promoting human motion digital twin. This work proposes an inertial motion capture system for human motion digital twin. On the basis of the developed system, this work focuses on the attitude calculation method of adaptive extended Kalman filter and the attitude calibration method based on kinematic constraints. The proposed system achieves an average root mean square error of 4.7° in estimating orientations in comparison with an optical motion capture system. Experimental results show a great correlation (92.5%) between the proposed system and the optical system, and satisfy the requirement that over 95% of data points reside within the consistency interval. The abilities of the proposed system focus on human-centric applications based on the integration of human, cyber system, and physical system, such as motion monitoring, human–robot teleoperation.

Keywords Motion perception · Human motion perception · Inertial motion capture · Adaptive extended Kalman filter

2.1 Introduction

This chapter focus on the key technologies of human motion perception and computation, particularly pose estimation of sensor nodes and human motion calibration. Firstly, an extended Kalman filter algorithm based on adaptive error estimation for pose estimation is proposed. This algorithm employs quaternions to represent orientations in multi-dimensional space and transforms complex three-dimensional rotations into an optimization problem solved through a nonlinear least squares method. Notably, the algorithm uses an adaptive strategy that dynamically adjusts the process

H. Zhou et al., *Revitalizing Human-Robot Interaction*, Advanced Topics in Science and Technology in China 3, https://doi.org/10.1007/978-981-96-6139-8_2

noise covariance matrix based on real-time data. This effectively compensates for modeling errors caused by sensor noise, external disturbances, and other factors, thus enhancing the accuracy and stability of the pose estimation. At the same time, a novel human motion calibration algorithm based on kinematic constraints is proposed for human motion reconstruction. It employs three specific postures to accurately establish the relationship between the sensor coordinate system and those of various body parts. This provides a precise attitude reference for the collaborative work between intelligent robots and operators. Additionally, by considering kinematic constraints, the algorithm significantly reduces calibration errors, thereby enhancing the accuracy and reliability of human motion capture. This method ensures that the attitude information from a single sensor node is accurately mapped to each part of the human body, and it is helpful for facilitating more precise human motion reconstruction.

To validate the effectiveness of the proposed algorithms, both the pose estimation and calibration algorithms were integrated into the self-developed inertial motion capture system. Specified experimental tasks were designed to assess the pose estimation capabilities of individual inertial sensor nodes and the overall human motion reconstruction performance. Experimental results demonstrate that the proposed algorithms significantly improve pose estimation accuracy, stability, and the accuracy of human motion reconstruction compared to traditional methods.

2.2 Motion Perception Based on IMUs

2.2.1 Initial Orientation Solution

In this section, the motion data of the IMU node in the stationary state is used to solve the initial orientation of the sensing node. The measured acceleration and the measured magnetic field of the IMU are put into the vector y_m, showing in the sensor coordinate system. When the sensor is in the stationary state, the theoretical acceleration and the theoretical magnetic field of the IMU are concatenated as vector y_g, showing in the geographic coordinate system. Assuming that the attitude quaternion of the IMU node is known to be q, y_g is converted with the aid of q to y_{est} of the sensor coordinate system, where y_{est} represents the estimated value of the combined vector of the acceleration vector and the magnetic field vector in the sensor coordinate system. Due to the measurement and modeling error, there is error between the measured value y_m and the estimated value y_{est}. Taking the attitude quaternion of the IMU, q as the variable, the error function $d(q)$ is constructed and calculated using the equation,

$$d(q) = y_m - y_{est} \tag{2.1}$$

To minimize the error function, q needs to satisfy the following iterative function,

$$q^{+} = q^{-} - \left(J\left(q^{-}\right)^{T} J\left(q^{-}\right)\right)^{-1} J\left(q^{-}\right)^{T} d\left(q^{-}\right) \tag{2.2}$$

where q^{+} represents the estimated quaternion after iteration, and q^{-} is the quaternion before iteration, and $d(q^{-})$ is the error function associated with q^{-}, and $J\left(q^{-}\right)$ is the Jacobi matrix of q^{-}, and the $(\cdot)^{T}$ represents the transpose operation, and $(\cdot)^{-1}$ represents the inverse operation. The Gauss–Newton algorithm is used to solve the above problem iteratively to find the optimal value of the attitude quaternion q, i.e., to obtain the initial orientation of the IMU node.

2.2.2 Adaptive Extended Kalman Filter

Extended Kalman filter updates the estimated attitude quaternion by multi-sensor data fusion, based on the initial orientation quaternion solved using the previous algorithm. In this chapter, the angular velocity $\omega(t)$ and quaternion $q(t)$ are selected as the state variables of the system, and the quaternion differential equation is introduced. At the same time, the coefficient of the first-order model $\tau(t)$ and the process noise $w(t)$ are introduced to represent the differential relationship between the angular velocity $\omega(t)$ and the angular acceleration $\dot{\omega}(t)$ by the first-order model, and the system state equation is established as follows,

$$\dot{X}(t) = \begin{bmatrix} \dot{\omega}(t) \\ \dot{q}(t) \end{bmatrix} = \begin{bmatrix} -\frac{1}{\tau(t)}\omega(t) + \frac{1}{\tau(t)}w(t) \\ \frac{1}{2}q(t) \otimes \left[\, 0 \ \omega(t) \,\right] \end{bmatrix} \tag{2.3}$$

After that, the process nonlinear transition function $f[\cdot]$ is introduced. The above nonlinear function is linearized around the optimal state estimate $\hat{X}(t)$. When it is in the optimal estimate state, it is expanded into a Taylor series and its approximate value is taken using the equation,

$$\dot{X}(t) = f[X(t), t]|_{X(t)=\hat{X}(t)} + \frac{\partial f[X(t), t]}{\partial X(t)}|_{X(t)=\hat{X}(t)}\left(X(t) - \hat{X}(t)\right) + w(t) \tag{2.4}$$

Then, $e(X(t))$ is defined as the increment between the $X(t)$ and $\hat{X}(t)$. The differential equation can be discretized as,

$$e(X_k) = \Phi_{k,k-1} e(X_{k-1}) + w_{k-1} \tag{2.5}$$

$$\Phi_{k,k-1} \simeq I + F_{k-1}\Delta t \tag{2.6}$$

where $\Phi_{k,k-1}$ is the transformation matrix, w_{k-1} represents the discrete process noise at time step $k-1$, and Δt is a small sampling interval used for the discretization process. Q_{k-1} and P_{k-1} represent the discrete process noise covariance matrix and the posterior covariance at time step $k-1$, respectively. Hence the prior estimate covariance $P_{k,k-1}$ can be calculated using the following equation,

$$P_{k,k-1} = \Phi_{k,k-1} P_{k-1} \Phi_{k,k-1}^T + Q_{k-1} \tag{2.7}$$

At the same time, the state output are six-dimensional vectors, consisting of the measured acceleration a_k and the measured magnetic field m_k at time step k. The observations can be calculated using the equation,

$$Z_k = \begin{bmatrix} a_k \\ m_k \end{bmatrix} \tag{2.8}$$

Combining the system state variables, the following observation equations can be obtained, which is,

$$Z_k = H_k X_k + v_k = \begin{bmatrix} 0 & g^q \\ 0 & m_g^q \end{bmatrix} X_k + v_k \tag{2.9}$$

where H_k consists of the gravity vector g^q and the local magnetic vector m_g^q, and v_k represents the measurement noise at time step k. The innovation of state output can be calculated using the equation,

$$\delta_k = Z_k - H_{k-1} X_{k-1} \tag{2.10}$$

The Kalman gain K_k of the system is expressed as follows.

$$K_k = P_{k,k-1} H_k^T \left[H_k P_{k,k-1} H_k^T + R_k \right]^{-1} \tag{2.11}$$

$$P_k = (I - K_k H_k) P_{k,k-1} (I - K_k H_k)^T + K_k R_k K_k^T \tag{2.12}$$

where R_k is the measurement noise covariance matrix, and P_k is the posterior covariance of the new estimate, and I is the identity matrix.

Measurement noise covariance matrix R_k consists of two main components, i.e., IMU measurement noise R_k^{mea} and the covariance matrix related to the acceleration modeling error R_k^{mod}. In this chapter, an adaptive strategy is used to estimate R_k^{mod} which is specified as follows. When the actual measurement error of the model exceeds the estimation error, it indicates that the measurement noise covariance matrix R_k can no longer accurately reflect the dynamically changing error disturbances. Therefore, an adaptive strategy needs to be introduced to estimate the error, which can be calculated using the equation,

$$tr\left(\delta_k \delta_k^T\right) \geq tr\left(H_k P_{k,k-1} H_k^T + R_k^{mea} + \widehat{R_k^{mod}}\right) \tag{2.13}$$

where $\delta_k \delta_k^T$ represents the actual error of the model, and $H_k P_{k,k-1} H_k^T + R_k^{mea} + \widehat{R_k^{mod}}$ represents the computed error, and $\widehat{R_k^{mod}}$ represents the theoretical value of R_k^{mod}. To obtain the estimated value E_k^{mod} of the modelling error, the modelling error is estimated using the error based on the sliding time window, i.e., using the average value of the time window length τ instead of the error at time k. E_k^{mod} can be calculated using the equation,

$$E_k^{mod} = diag\left(\frac{1}{\tau}\sum_{j=k-\frac{\tau}{2}}^{k+\frac{\tau}{2}} \delta_j \delta_j^T - H_k P_{k,k-1} H_k^T - R_k^{mea}\right) \tag{2.14}$$

where E_k^{mod} represents the tuned modeling error innovation covariance, and $diag(\cdot)$ represents the diagonal operation.

Meanwhile, in the actual measurement process, the noise parameter is a dynamic variable. In this chapter, the forgetting factor b is introduced to emphasise the role of recent data and reduce the weight of old data. Through the weighted calculation method, R_k can be calculated by the following equation,

$$R_k = \frac{1-b}{1-b^{k+1}}\left(\delta_k \delta_k^T - H_k P_{k,k-1} H_k^T\right) + \left(1 - \frac{1-b}{1-b^{k+1}}\right)\left(R_k^{mea} + E_k^{mod}\right) \tag{2.15}$$

In the motion perception, the initial orientation vector of the IMU node is first solved iteratively using the Gauss–Newton algorithm. Then the extended Kalman filter algorithm based on adaptive error estimation is employed to iteratively update the attitude quaternion to obtain the motion attitude of the IMU node.

2.3 Human Motion Calibration and Reconstruction

Human motion reconstruction is a critical research issue in the human–robot teleoperation. It involves capturing and analyzing the movements of various parts of the human body to understand and simulate the mechanisms of human motion. This study employs multi-node inertial sensors to collect motion data and uses this data to reconstruct human motion, which involves complex data processing and computational models.

This section focuses on how to use these inertial motion data to reconstruct the motion of the human body. Based on the anatomical structure and kinematic characteristics of the human body, this research utilizes a simplified skeletal model to represent the human body. This model abstracts the human skeleton and joints according to the real hierarchical structure of the human skeleton. The bones of each limb segment are represented as vectors of a certain length, and the joints connecting limb segments are represented according to the degree of freedom, such as hinge joints (Fig. 2.1).

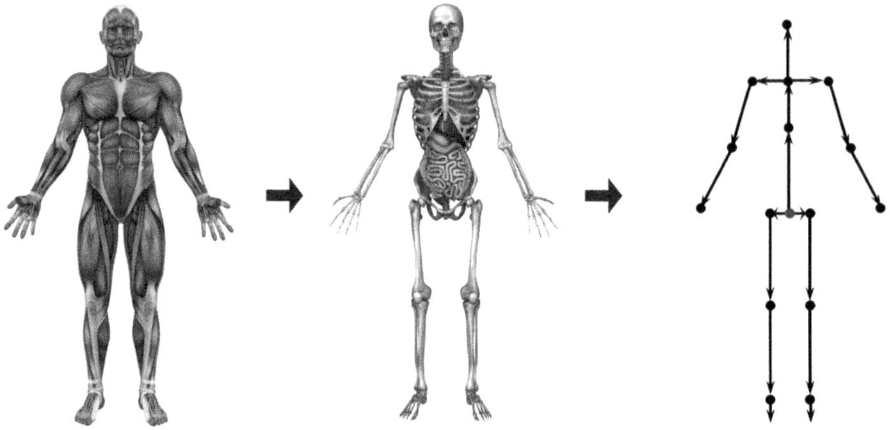

Fig. 2.1 Simplified vector model of the human skeleton

During the model construction, the midpoint of the two hip joints is used as the root node of the entire model. This design allows the model to better reflect the dynamic changes of the human body's center of gravity. The construction of limb segment vectors adheres to the natural order of human movement, starting from the root node, with the upper limbs arranged from bottom to top and the lower limbs from top to bottom. The start point of each vector is defined as the parent node, and the end point as the child node. Through this parent–child node relationship, a simplified vector model of the human skeleton is formed.

This chapter introduces a human motion calibration method that integrates human kinematic constraints. This method specifically defines three particular calibration poses and uses motion data collected through these poses to estimate the directional axes of the local body coordinate systems (Table 2.1). Furthermore, it utilizes the principles of human kinematics to correct these axes' orientations, ensuring that the posture data accurately reflects the body posture. The calibration poses are described in detail as follows.

Table 2.1 Description of the calibration movements

Movement	Description
Pose 1	Stand with torso upright, dual arms hanging naturally, palms faced to body, feet shoulder-width apart and being parallel to each other
Pose 2	Sit down with trunk forwards inclined, and rest elbows on the table and swinging shanks around the knees while keeping thighs lying on the chair
Pose 3	Sit down with palms facing down and raising arms out to the sides until arms are slightly above parallel to the floor, and keep thighs parallel to the floor and thighs parallel to the ground and shanks perpendicular to the floor

Calibration Pose 1: The participant is required to stand upright, keeping the torso straight, arms naturally hanging at the sides of the body, and palms facing the body. The feet should be shoulder-width apart, with the soles parallel and pointing forward. This pose helps to accurately define the body's vertical and horizontal baselines.

Calibration Pose 2: The participant needs to maintain a seated position, lean the upper body forward, and place the elbows and forearms on the tabletop, ensuring that the elbow spacing matches the shoulder width, with the forearms kept parallel. The thighs should be flat on the chair, parallel to the ground while swinging the lower legs forward and backward around the knees. This pose aids in accurately correcting the data for knee joints and lower limb movements.

Calibration Pose 3: While seated, the participant raises the arms to the sides of the body, with palms facing down and parallel to the ground. The thighs remain flat on the chair, parallel to the ground, with the lower legs at a 90° to the floor, feet shoulder-width apart, and soles parallel and facing forward. This pose helps to accurately define the range of motion for the upper limbs and shoulder joints.

The limb segment posture transformation matrix is defined as $M_{trans} = \left[X_{trans}\ Y_{trans}\ Z_{trans} \right]$. Combining the motion data of different calibration postures, the anatomical axis pointing of each limb segment is determined, i.e., X_{trans}, Y_{trans}, Z_{trans}. In this section, the attitude transformation matrix solving method is illustrated by solving the attitude transformation matrix of the thoracic spine segment and the attitude transformation matrix of the calf segment as an example. The determination of the posture transformation matrix for the thoracic spine segment requires the use of calibration motion data from calibration pose 1 and calibration pose 2. When the human body is in a stationary upright state (calibration pose 1), the direction of the acceleration vector of the IMU node can be considered as the pointing of the coordinate axis of the corresponding limb segment. Thus, the Y-axis of the joint coordinate system of the thoracic spine segment is determined on the basis of the acceleration of the inertial sensing node fixed to the spine segment of the human body, calculated as follows,

$$Y_{spine-trans} = \left[a_{spine-x1}\ a_{spine-y1}\ a_{spine-z1} \right]^T \tag{2.16}$$

where $\left[a_{spine-x1}\ a_{spine-y1}\ a_{spine-z1} \right]$ is the acceleration vector measured by the IMU node fixed in the spine segment in calibration pose 1. Unlike the state of the spine in the upright posture, the spine remains in an inclined state in calibration posture 2, i.e., the Y-axis in the coordinate system of the thoracic vertebral segments is along the inclined state. Using the Y-axis vectors of the different states to determine the plane in which they are located. By considering the Z-axis vector as the normal vector to this plane, the Z-axis of the coordinate system of the thoracic spine segment can be determined, which can be calculated by the following equation,

$$Z_{spine-trans} = \frac{Y_{spine-trans} \times normal\left(\left[\, a_{spine-x2}\ a_{spine-y2}\ a_{spine-z2} \,\right]^{T}\right)}{\left|Y_{spine-trans} \times normal\left(\left[\, a_{spine-x2}\ a_{spine-y2}\ a_{spine-z2} \,\right]^{T}\right)\right|} \tag{2.17}$$

where $\left[\, a_{spine-x2}\ a_{spine-y2}\ a_{spine-z2} \,\right]$ represents the spine segment inertial sensing node acceleration vector acquired in calibration pose 2, and $normal(\cdot)$ represents the normalization operation. Based on the solved Y-axis and Z-axis, the X-axis of the coordinate system of the thoracic spine segment can be determined, which can be calculated by the following equation,

$$X_{spine-trans} = \frac{Y_{spine-trans} \times Z_{spine-trans}}{\left|Y_{spine-trans} \times Z_{spine-trans}\right|} \tag{2.18}$$

So far, the three sub-vectors of the posture transformation matrix of the thoracic vertebral segment are established.

Soft tissue artifacts such as shading and irregular calibration postures can lead to errors in the calibration data when performing calibration postures. In this section, kinematic constraints and human motion calibration algorithms are combined with human kinematic constraints. Firstly, principal component analysis is applied to determine the primary axis of rotation and try to avoid the introduction of rotation errors in other axes. Secondly, the pose transformation matrix is corrected by compensating for errors in secondary joint angles. For example, in calibration posture 3, there is only abduction and adduction motion of the shoulder joint, and an irregular calibration posture may produce secondary error joint angles of internal and external rotation and flexion and extension. The error of secondary joint angles is compensated on the basis of the initial orientation transformation matrix ascertained from the calibration pose.

With the human body attitude calibration, the attitude transformation matrix between the inertial sensing nodes and the main limb segments of the human body can be obtained. Based on this transformation matrix, the human motion reconstruction algorithm transforms the solved inertial sensing node's pose into limb segment poses. The human motion is reconstructed based on human kinematics and skeleton modeling. The real-time attitude quaternions in the sensor coordinate system of each node are $q_s = \left[\, q_s^0\ q_s^1\ q_s^2\ q_s^3 \,\right]$. The attitude transformation matrix $M_{trans} = \left[\, X_{trans}\ Y_{trans}\ Z_{trans} \,\right]$ obtained via attitude calibration is transformed into the quaternion form $q_{trans} = \left[\, q_{trans}^0\ q_{trans}^1\ q_{trans}^2\ q_{trans}^3 \,\right]$.

According to the quaternion arithmetic rule, the real-time limb segment attitude under the human body coordinate system is obtained as q_b, which can be calculated by the following simple equation,

$$q_b = q_s \otimes q_{trans} \tag{2.19}$$

By performing the above calculations on human limb segments one by one, the real-time postures of the main limb segments of the human body can be obtained to reconstruct human motion.

2.4 Experiment and Results

During the experimental validation of the algorithm, researchers typically adopt a stepwise strategy. Firstly, the algorithm undergoes preliminary evaluation in a laboratory environment. This step facilitates the swift identification and resolution of fundamental technical issues. Secondly, the experimental scope is expanded to encompass diverse user groups, a wide range of movements, and various environmental conditions. This step aims to assess the performance of the algorithm under conditions approximating real-world applications. It necessitates the collection and analysis of extensive data. Through comprehensive and systematic testing, researchers can evaluate the feasibility and effectiveness of the algorithm in practical applications. The architecture of the inertial motion capture system-based application is shown in Fig. 2.2.

This section's experiments aim to verify and demonstrate the effectiveness and accuracy of the human motion calibration method and motion reconstruction method proposed in this study. There are two experiments, i.e., pose estimation experiments utilizing single inertial sensor node and human motion reconstruction experiments using multiple inertial sensor nodes. The developed system consists of four main parts, i.e., IMU nodes, a hub node, a router and a personal computer (PC). An IMU node incorporates a tri-axis accelerometer, a tri-axis gyroscope, and a tri-axis magnetometer. When gathering and handling motion data from IMU nodes, the

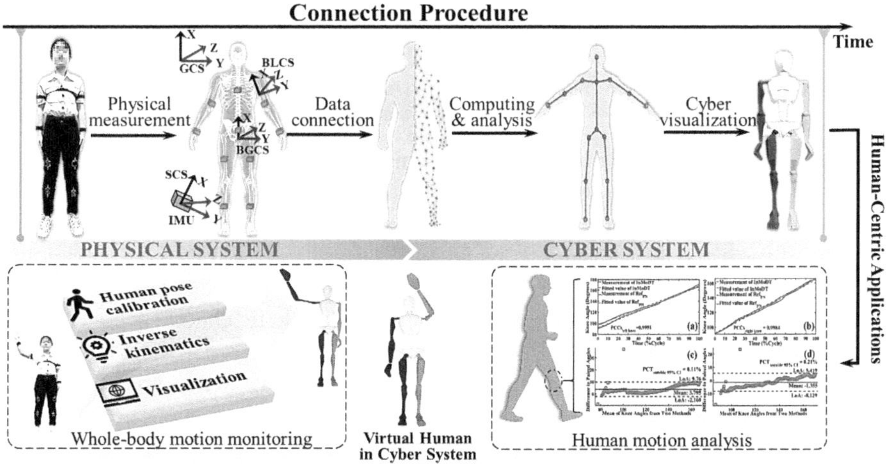

Fig. 2.2 Architecture of the inertial motion capture system-based application

microcontroller unit (MCU) transmits the data to the portable hub node. The hub node retrieves all the motion data and wirelessly sends the processed data stream to a PC. It enables the option of adding or removing IMU nodes as required. The Wi-Fi module of the hub node converts the data from the Serial Peripheral Interface (SPI) into Transmission Control Protocol (TCP) data packets. The router reads the network address in the header of the motion data packet and directs the packet to the specific PC through TCP.

First, the design of the single inertial sensor node pose estimation experiment aims to assess and validate the performance of the pose estimation algorithm with single sensor data input. Optical motion capture systems, which provide high-precision spatial positioning data, serve as the references standard. It is critical for comparing and evaluating the data processing algorithms of inertial sensor nodes. Additionally, by comparing the effects and accuracy of this study's algorithm with commercial inertial algorithms processing the same sensor data, the accuracy of this study's algorithm is verified.

Second, the human motion reconstruction experiment based on multiple inertial sensor nodes further expands the scope of research, involving more complex data integration and processing scenarios. The design of this experiment reflects the algorithm's processing capabilities and accuracy in complex movement and multi-node data environments. By using both self-developed and commercial motion capture devices, the experiment assesses not only the algorithm's efficacy in managing multi-source data but also its competitiveness against current technologies. Moreover, the experimental setup of static and dynamic tasks is intended to simulate different real-world application scenarios, ranging from simple static poses to complex dynamic walking, ensuring the technology can adapt to various types of movement.

Finally, through the above experiments, it can be ensured that the algorithms proposed in this study meet the high standards of accuracy and flexibility required in practical applications of motion capture.

2.4.1 Experiment of Single IMU Node

The single inertial sensing node attitude solving experiment takes the optical motion capture system as the reference standard and introduces a variety of inertial sensing node attitude solving comparison algorithms. The attitude solving experiment consists of two sets of experiments, which are experimental task 1 and experimental task 2.

In experimental task 1, the experimenters fixed a single IMU node at the center of the plastic holder, while four optical reflective markers were fixed to the corners of the plastic holder. Participant used straps to bind the plastic holder to the subject's feet and ensured that the plastic holder and the optical reflective markers were not obscured. This did not interfere with the experimental participant's normal walking, as shown in Fig. 2.3. During the experiment preparation stage, the participant is required to stand in the designated position according to the experiment requirements. The static

Fig. 2.3 Experimental setup for attitude solving of a single inertial sensing node

upright standing posture was maintained for a number of seconds to synchronize and calibrate the two motion capture systems. The Qualisys 3D motion acquisition and analysis system and the inertial sensing node recorded the experimental data synchronously. In experimental task 2, the commercial inertial motion capture algorithm is compared. The experimenters stacked and fixed the self-developed IMU node and the commercial IMU in the center of the plastic holder. This step aims to make the sensor axes of the two inertial sensing nodes pointing in the same direction. At the same time, four optical reflective markers were fixed at the corners of the plastic holder. During the experiment, the participant held the plastic stand, rotating it around the sensor coordinate system's X, Y, Z, and arbitrary axes at a steady pace.

In this experiment, the Qualisys motion acquisition and analysis system is employed as the standard system. A variety of attitude solving algorithms based on inertial motion data [1–8] are compared for their efficacy to solve the attitude of inertial sensing nodes.

Zhang et al. proposed an adaptive zero-velocity update (ZUPT) algorithm that utilizes an additional chest accelerometer to detect the stationary phases of pedestrians [1]. It mitigates the position drift issues caused by cumulative errors in low-cost MEMS sensors during pedestrian tracking. Marantos et al. introduced an adaptive complementary filter for UAV state estimation suitable for high-dynamic flights and GPS-constrained environments, which enhances the accuracy and reliability of position and attitude estimation by integrating multi-sensor data [2]. Ligorio et al. presented a new linear Kalman Filtering-based estimators which address the tilt angles estimation problem (i.e. the pitch and roll angles, or the attitude) as a source separation technique applied to the accelerometer signal [3]. Wu et al. developed a multi-sensor fusion attitude estimation algorithm using a linear complementary filter and gradient descent algorithm to process the quaternion-based observation model [4]. This method significantly reduces computational complexity by simplifying the observation model to a linear form, improving the efficiency and accuracy of estimations compared to traditional nonlinear methods. Chao et al. showed a minimum settings calibration method for low-cost tri-axial IMUs and magnetometers [5]. They utilized the ellipsoid model to calibrate the tri-axial accelerometer and magnetometer preliminarily, then completed the calibration work with at least two reference angular positions, and proved that two reference angular positions are minimum requirements in the calibration progress. Javed et al. proposed a method for accurately tracking vehicle orientation in harsh environments using a novel filtering framework based on a linear Kalman filter, utilizing sensor constraints to eliminate uncertainties caused by acceleration, thus enhancing positioning accuracy in dynamic scenarios without additional sensor information [6].

In the experiment, the attitude angles solved by the optical motion capture system are considered standard, and the Root Mean Square Error (RMSE) of the Euler angles is served as a metric to indicate the accuracy of the algorithms,

$$ang_{rmse} = \sqrt{\frac{1}{n}\sum_{1}^{n}(I(t) - Q(t))^2} \tag{2.20}$$

where ang_{rmse} represents the attitude angle RMSE, and $I(t)$ represents the attitude angle obtained by applying the proposed algorithm at the moment t, and $Q(t)$ represents the attitude angle based on the optical motion capture system at the moment t, and n is the discrete time of an experiment.

Figure 2.4 shows the RMSEs of the orientation solving algorithms for the inertial sensing node, including the solving accuracies of the computed roll angle, pitch angle, yaw angle, and the average of the three attitude angles. From the experimental results, it can be seen that the algorithm in this chapter using the adaptive strategy outperforms the other algorithms. The adaptive error estimation strategy effectively reduces the Euler angle RMSE from 3.486°to 1.993°. The commercial inertial attitude solving algorithm [8] performs poorly at 2.358°. Although the algorithm proposed by Marantos achieves full state estimation, it is possible that its target is a flying vehicle, which differs from the human motion state estimation application goals and scenarios in this chapter. The average value of the RMSE of the Euler angles solved by this algorithm is 7.407°.

Table 2.2 demonstrates the variations of inertial sensing node poses solved by the algorithms presented in this chapter and the commercial inertial pose solving algorithm across four different rotational motions. The difference between the two is measured by RMSE. When the rotation of the inertial sensing node around the X-axis and arbitrary axis of the sensor coordinate system, two algorithms solve the attitude quaternions with similar trends. In contrast, when the inertial sensing node is rotated around the Y-axis and the Z-axis, the two attitude quaternions show almost exactly opposite values at some moments of motion.

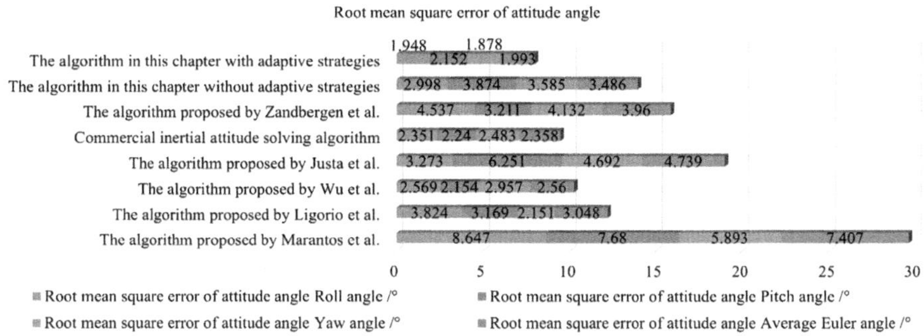

Fig. 2.4 RMSEs of orientation estimation results by different algorithms

Table 2.2 Root mean square error of solved poses of the variations between the commercial algorithm and the proposed algorithm

Algorithm	Root mean square error of attitude angle			
	q_0	q_1	q_2	q_3
X axis	0.0065	0.0614	0.0619	0.0070
Y axis	0.0062	0.0052	0.0556	0.0079
Z axis	0.0364	0.0341	0.0753	0.0756
Arbitrary axis	0.0066	0.0118	0.0113	0.0148

2.4.2 Experiment of Human Motion Reconstruction

The human motion reconstruction experiment comprises two tasks, i.e., experimental task 1 and experimental task 2. Experimental task 1 was a static pose reconstruction experiment. The participant wore both the self-developed inertial motion capture device and the Qualisys optical motion capture device (Fig. 2.5). In the control group, the same participant wore both the commercial inertial motion capture device, Perception Neuron 2.0, and the Qualisys optical motion capture device. In each set of experiments, the participant needed to ensure that the wearing positions of the two sets of motion capture devices did not interfere with each other, in order not to affect the normal operation of each other's systems. During the preparation phase, the participant wearing the two sets of motion capture devices completed the calibration and alignment as per the system requirements. During the experimental phase, the participant performed four sets of leg flexion and extension movements, i.e., holding the knee flexion and extension angles of 90°, 120°, 150°, and 180° sequentially. One experiment consisted of the leg changing from knee flexion and extension of 90° to 180° and then from 180° to 90°. Experimental task 2 was a dynamic motion reconstruction experiment, and the experimental setup was the same as experimental task 1, with different experimental content than the former. The participant wearing two sets of motion capture devices were required to walk in the experimental site at

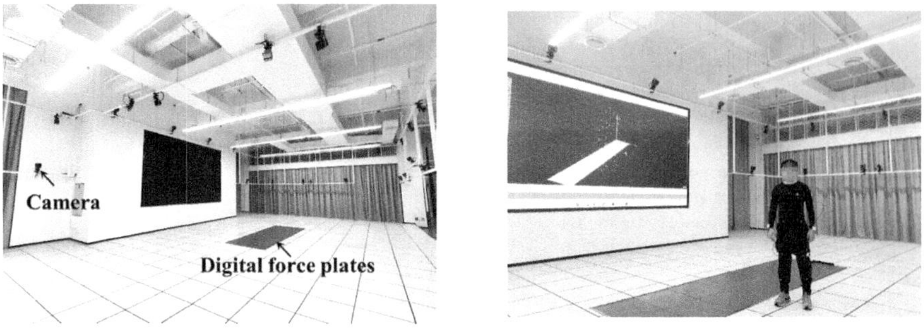

(a) Motion capture experimental site (b) Experimental setup for static postures and dynamic movements

Fig. 2.5 Experimental site and experimental setup for human motion reconstruction

a normal walking speed in accordance with a designed straight-line route, and one experiment consisted of completing a back-and-forth walk.

(1) **Static posture reconstruction experiment**

In the static pose reconstruction experiments, the inertial motion capture system and the optical motion capture system solved the knee flexion and extension angles using the collected static pose motion data. The calculation results compare joint angles solved based on the optical system as the reference, and the mean absolute error and the RMSE are used as the comparison metrics. The experimental results are shown in Fig. 2.6. When the extended Kalman filter algorithm based on adaptive error estimation and the motion calibration algorithm based on kinematic constraints are integrated in the self-developed inertial motion capture system, its reconstructed static joint angle has a RMSE of 3.889°. The average absolute error of the static pose joint angles solved by the commercial inertial motion capture system was the RMSE was 6.982°. Results indicate that the self-developed inertial motion capture system slightly outperforms the commercial system in static pose reconstruction. Meanwhile, when the pose solving algorithm without kinematic constraints is applied to solve the joint angles of static poses, the average absolute error and the RMSE of the corresponding joint angles are both larger than 10°, indicating that the application of kinematic constraints helps to improve the accuracy of the static pose. When solving the joint angles using the self-developed inertial motion capture system and the commercial inertial motion capture system, the error of the static joint angle of the left leg was significantly larger than the static joint angle of the right leg. This may be explained by the skin tissue artefacts shadowing and the offset of sensor wear affecting the measurement accuracy of the motion capture system.

Meanwhile, correlation analysis and agreement analysis of the self-developed system and the commercial system are done in this experiment.

Correlation analysis: Joint angles acquired from two sets of data are assessed to analyze the correlation between the self-developed system and the commercial

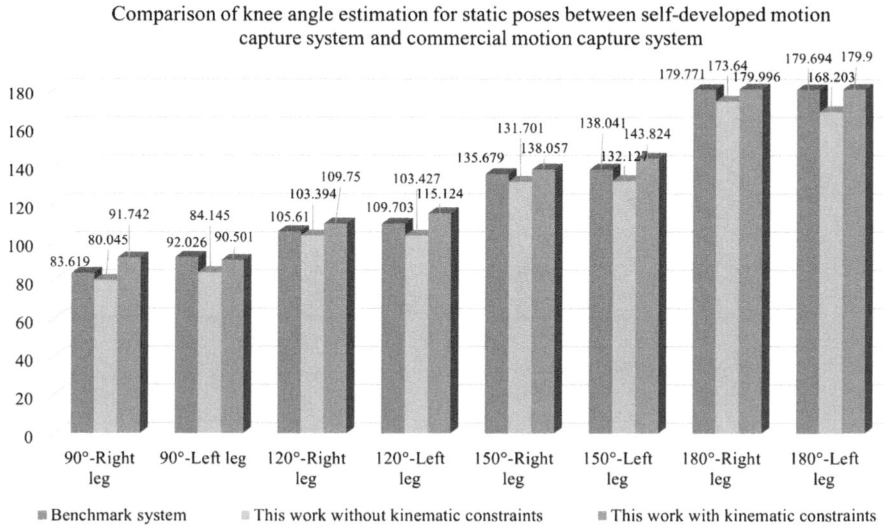

Fig. 2.6 Comparison of knee angle estimation for static poses between self-developed motion capture system and commercial motion capture system

system during dynamic movements for four experimental tests respectively. Each test of "Standing" and "Sitting" is completed with the same experimental setup. The scattered plot and linear fitting lines associated with joint angles for correlation analysis are shown in Fig. 2.7. Pearson correlation coefficients (PCCs) are evaluated between each test of dynamic motion data, as reported in Table 2.3. The maximum value of PCCs is 0.9993 and the minimum value is 0.9647, which is larger than 0.95. It reveals a strong positive correlation between the self-developed system and the commercial system. The observed measurement of the commercial system is related human motion. Thus, it concludes that the self-developed system is also being affected in a similar manner due to the strong correlation. Besides, it infers that the self-developed system's measurement shows a small offset of joint angles waveforms without noticeable distortion in comparison with the commercial system for motion measurement.

Agreement analysis: To assess the agreement between measurements obtained from the self-developed system and the commercial system, Bland–Altman analysis is performed to plot the difference between the two paired measurement results during dynamic movements. If the differences are normally distributed, the difference will lie between $d - 1.96s$ and $d + 1.96s$ while d is the mean difference and s is the standard deviation of the differences. These limits form the confidence intervals for Bland–Altman analysis. Figure 2.8 represents the agreement analysis results between measurement obtained from two systems, using knee joint angles of lower limb motion during four tests. The dotted lines represent the upper and the lower

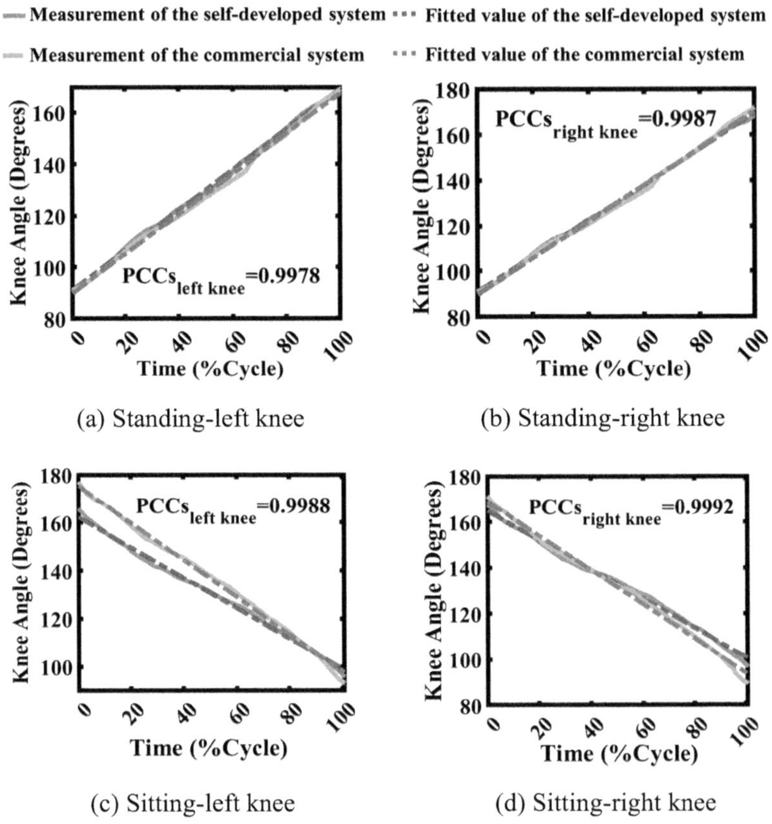

Fig. 2.7 Correlation analysis between different inertial motion capture devices for dynamic movements

Table 2.3 PCCs of different inertial motion capture devices for standing and sitting movements

Standing	PCCs[a] for left knee	PCCs for right knee	Sitting	PCCs for left knee	PCCs for right knee
Test 1	0.9978	0.9987	Test 1	0.9991	0.9984
Test 2	0.9966	**0.9993**	Test 2	**0.9647**	0.9779
Test 3	0.9967	0.9972	Test 3	0.9988	0.9992
Test 4	0.9689	0.9788	Test 4	0.9996	0.9991

[a] *PCCs* Pearson correlation coefficients

limits of agreement (LoA), and the solid lines show the mean value of the differences. Table 2.4 shows the results of the agreement analysis, including each test of "Standing" and "Sitting". The percentage of out of the confidence interval varies from 0% to 4.85%, which is less than 5%. It allows more than 95% certainty in

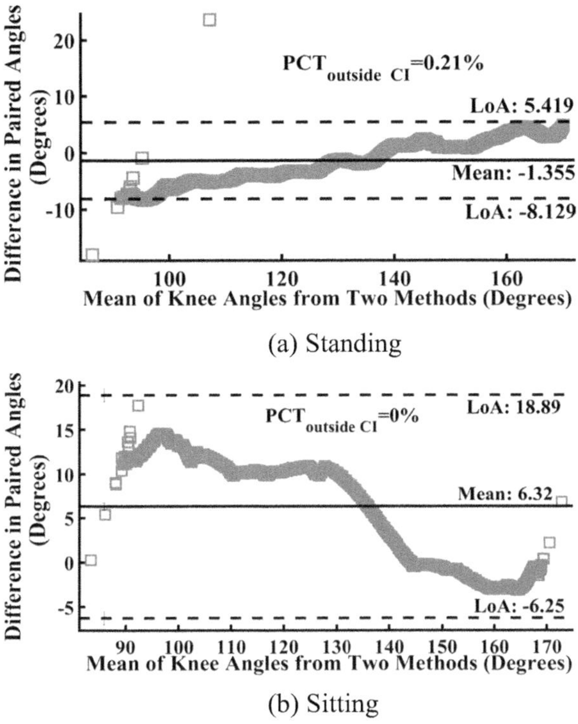

Fig. 2.8 Bland–Altman analysis between different inertial motion capture devices

sample variables, which indicates the agreement between the self-developed system and the benchmark system.

(B) **Dynamic motion reconstruction experiment**

The spinal flexion/extension angle, spinal rotation angle, shoulder flexion/extension angle, elbow flexion/extension angle, hip abduction/adduction angle, and knee flexion/extension angle were selected for analysis in this experiment. In this section, the RMSE of joint angles, Pearson correlation coefficient, and agreement analysis were utilized as evaluation metrics. RMSE is intended to reflect the deviation between the estimated and standardized values. The Pearson correlation coefficient is often applied to measure the correlation between two variables. The absolute Pearson correlation coefficient value between 0–30% indicates a weak correlation, 30%-70% suggests a moderate correlation, and 70–100% denotes a strong correlation between the two variables [10]. The purpose of the agreement analysis test is to compare the consistency of the results obtained by different methods. Bland–Altman agreement analysis was used in this chapter.

Figure 2.9 shows the RMSEs of the joint angles solved by the self-developed inertial motion capture system based on the motion calibration algorithm with and without kinematic constraints. Consistent with the static pose joint angle solving results, the errors of the solved joint angles of the algorithms without kinematic

Table 2.4 Bland–Altman analysis between different inertial motion capture devices for standing and sitting movements

Test movement	LoA[a]	Mean[b]	PCT[c] (%)
Standing-Test 1	[− 2.169, 9.76]	3.795	0.11
Standing-Test 2	[− 8.129, 5.419]	− 1.355	0.21
Standing-Test 3	[3.68, 11.14]	7.411	0.41
Standing-Test 4	[− 4.973, 3.725]	− 0.6242	4.85
Sitting-Test 1	[− 6.25, 18.89]	6.32	0
Sitting-Test 2	[− 2.816, 8.125]	2.655	1.31
Sitting-Test 3	[− 28.23, 5.019]	− 11.61	0.60
Sitting-Test 4	[− 4.215, 1.216]	− 1.5	3.94

[a] *LoA* Limits of agreement, computed as mean difference ±1.96SD (for homoscedastic data). [b] *Mean* The mean value of difference in paired angles. [c] *PCT* The percentage of out of the confidence interval

constraints are larger than those of the algorithms with kinematic constraints. In particular, the average RMSE of the joint angles solved by the former is 9.536°, whereas the latter achieves an average RMSE of 4.700°. In the attitude solution without kinematic constraints, the maximum value of the RMSE is 16.167° and the minimum value is 5.965°, with a large difference in accuracy between different joint angles. In the kinematic constraint-based joint angle solving, the maximum value of the RMSE is 5.998° and the minimum value is 2.524°, with similar solving accuracies among joint angles. Figure 2.10 reflects the Pearson correlation and agreement between the solved joint angles from the inertial motion capture system and the standard joint angles acquired by the optical system. The average Pearson's correlation coefficient between the calculated and standard values of joint angles is 92.5%, indicating a strong positive correlation. The minimum Pearson correlation coefficient is 84.5% for the right shoulder flexion/extension joint angle, and the correlation is generally low for upper extremity joints and relatively high for lower extremity joints. Similarly, the RMSE of the solved upper limb joint angles exceeds that of the lower limb joint angles. These findings may stem from the fact that upper limb joints possess more degrees of freedom and a broader range of motion. In the results of the agreement analysis experiments, an average of 2.190% of the joint angle values solved using the algorithms in this chapter fell outside the 95% consistency boundaries for 2.190% of the points. The highest percentage of data outside the consistency boundaries was 4.721% for the right shoulder flexion/extension angle. The lowest percentage of data outside the consistency boundaries was 0% for the right hip abduction/adduction angle and the left knee flexion/extension angle. The experimental results satisfy the requirement that over 95% of data points reside within the consistency interval.

In conclusion, the self-developed motion capture system, integrated with the algorithms presented in this chapter, demonstrates satisfactory performance in accuracy,

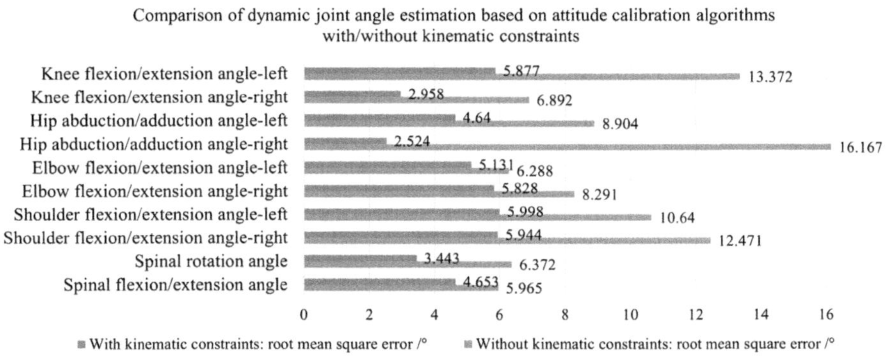

Fig. 2.9 Comparison of dynamic joint angle estimation based on attitude calibration algorithms with/without kinematic constraints

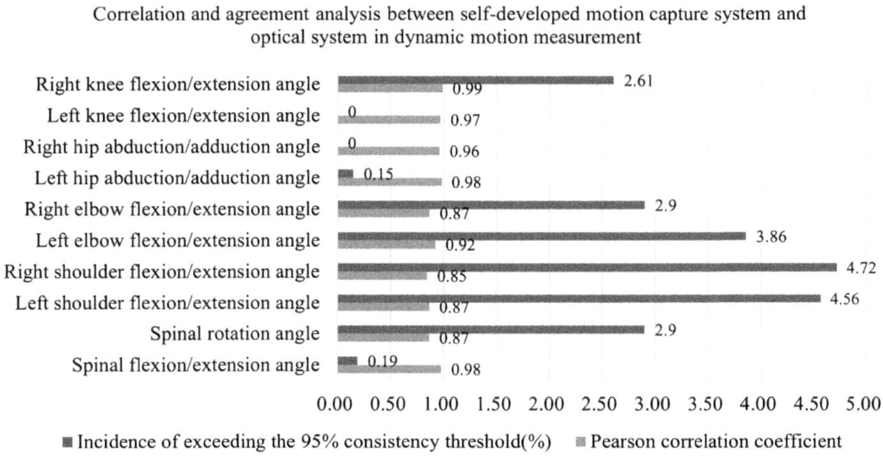

Fig. 2.10 Correlation and agreement analysis between self-developed motion capture system and optical system in dynamic motion measurement

relevance, and consistency, fulfilling the demands of daily applications [11, 12]. This system can provide data support for constructing human motion digital twin (Fig. 2.11).

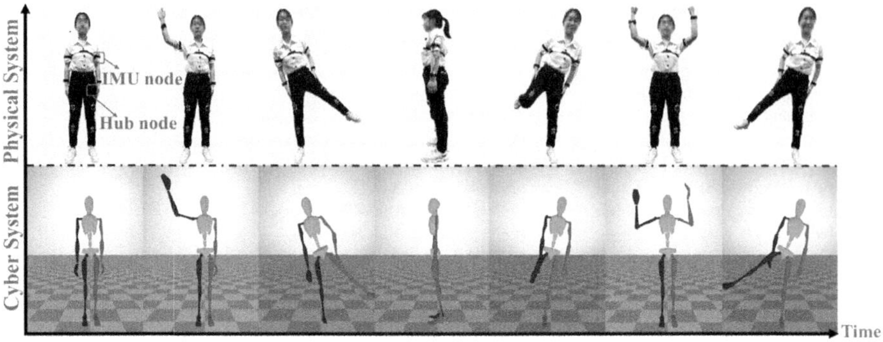

Fig. 2.11 Verification of human motion digital twin in whole body's movements

2.5 Conclusion

This chapter's research on IMU-based human motion perception and reconstruction aims to advance the field of smart manufacturing. It particularly emphasizes multi-sensor data fusion and accurate initial orientation vector solutions. The study employs an extended Kalman filter with adaptive error estimation to optimize the data processing process. Detailed discussions cover the calibration and reconstruction of human motion, focusing on the mapping sensor node data to human limb segments. It is crucial for robot control in smart manufacturing and for analyzing worker movements on automated assembly lines. Experiments validate the method's accuracy and reliability in single-node and multi-node sensor setups. These results establish a solid technical foundation for developing advanced human motion capture systems. For instance, it is possible to design more ergonomic automated workstations by capturing worker movements, thus reducing labor intensity and increasing work efficiency. In human–robot collaboration scenarios, precise human motion capture technology can facilitate safe interactions between robots and human, advancing smart manufacturing to a higher technical level.

References

1. Zhang R, Yang H, Höflinger F, Reindl LM (2017) Adaptive zero velocity update based on velocity classification for pedestrian tracking. IEEE Sens J 17(7):2137–2145
2. Marantos P, Koveos Y, Kyriakopoulos KJ (2016) UAV state estimation using adaptive complementary filters. IEEE Trans Control Syst Technol 24(4):1214–1226
3. Ligorio G, Sabatini AM (2015) A linear Kalman filtering-based approach for 3D orientation estimation from magnetic/inertial sensors. In: IEEE international conference on multisensor fusion and integration for intelligent systems. IEEE, San Diego, California, USA, pp 77–82
4. Wu J, Zhou Z, Fourati H, Li R, Liu M (2019) Generalized linear quaternion complementary filter for attitude estimation from multisensor observations: an optimization approach. IEEE Trans Autom Sci Eng 16(3):1330–1343

5. Chao C, Zhao J, Zhu J, Bessaad N (2021) Minimum settings calibration method for low-cost tri-axial IMU and magnetometer. Meas Sci Technol 33:025103
6. Javed MA, Tahir M, Ali K (2023) Attitude in motion: constraints aided accurate vehicle orientation tracking in harsh environment. In: IEEE transactions on industrial informatics, vol 19, no 5, pp 6450–6459
7. Justa J, Šmídl V, Hamáček A (2020) Fast AHRS filter for accelerometer, magnetometer, and gyroscope combination with separated sensor corrections. Sensors 20(14):3824
8. Zhu L, Zeng Y (2020) WISERING: research on intelligent motion-sensing interactive software and hardware integration system for multi-finger collaboration. In: IEEE international conference on electronics technology (ICET). IEEE, Chengdu, China, pp 142–146
9. Zandbergen MA, Reenalda J, van Middelaar RP et al (2022) Drift-free 3D orientation and displacement estimation for quasi-cyclical movements using one inertial measurement unit: application to running. Sensors 22(3):956
10. Choo CZY, Chow JY, Komar J (2022) Validation of the perception neuron system for fullbody motion capture. PLoS ONE 17(1):e0262730
11. Mcgrath T, Fineman R, Stirling L (2018) An auto-calibrating knee flexion-extension axis estimator using principal component analysis with inertial sensors. Sensors 18(6):1882
12. Bloomfield RA, Fennema MC, Mclsaac KA, Teeter MG (2019) Proposal and validation of a knee measurement system for patients with osteoarthritis. IEEE Trans Biomed Eng 66(2):319–326

Chapter 3
Human–Robot Teleoperation Based on IMUs

Abstract As a key enabling technology to advance human–computer interaction, human motion digital twin focuses on the behavior of human ontology in human–computer interaction. The combination of robot capabilities with human intelligence is gradually showing promising potentials. Teleoperation interaction is a mode of direct sharing of operation intention in human–robot interaction. However, due to the differences between the operator and the robot in movement, motion space and joint degrees of freedom, the teleoperation system cannot share the intention directly, accurate expression of human operation intention puts forward higher requirements for hardware resources and algorithm strategy. This work designs a generalized IoT-enabled telerobotic architecture by integrating human-motion-capture subsystem with robot-control subsystem. This work solves the motion trajectory of the end joint of the human body, according to the difference of human–robot motion characteristics and motion space, the motion-motion mapping strategy is designed to clarify the motion path planning of the teleoperation robot, which provides support for describing the human motion digital twin. As specific examples of application, the proposed architecture is applied to complete tasks based on dual-arm motion capture and mapping for telerobotics in laboratory and intercontinental scene.

Keywords Smart manufacturing · Human–robot teleoperation · Teleoperation interaction · Human–robot interaction · Collaborative robot

3.1 Introduction

With the growing prominence of teleoperation technology in critical fields such as telemedicine, disaster relief, and smart manufacturing, human motion capture and reconstruction technology has become crucial. This chapter focuses on human motion mapping interaction technology within teleoperation applications. Our research aims to enhance the efficiency and safety of teleoperation robots by capturing and mapping human motion to robots. This approach provides robust technical support for teleoperation and collaborative robots in smart manufacturing.

H. Zhou et al., *Revitalizing Human-Robot Interaction*, Advanced Topics in Science and Technology in China 3, https://doi.org/10.1007/978-981-96-6139-8_3

Our study begins with a human kinematics model, utilizing inertial motion capture technology to calculates the motion trajectory of joints. This process relies on understanding of human joint motion and the use of efficient, precise data processing algorithms. Then, we constructed skeletal models of human and robot, and calculated the operational range of workspaces between the operator and the robot. We designed a master–slave pose mapping strategy that fully considers the motion characteristics of the human arm. Thus, a motion path plan is determined for teleoperation. In the practical application, we implemented a teleoperation interaction application in the laboratory. A remote teleoperation interaction experiment spanning China and Sweden is conducted successfully. This provides a reliable technical foundation for tele-monitoring, teleoperation, and human–machine collaboration in smart manufacturing.

3.2 Human Motion Tracking

In the domain of teleoperation interaction, human motion serves as an intuitive reflection of the operator's intent, making the accurate capture and interpretation of human movement crucial for achieving efficient and natural control of robotic systems. By translating human gestures and motions into robot actions, the system facilitates seamless collaboration between humans and robots. This section focuses on developing an advanced tracking algorithm for the motion trajectory of the terminal joint of the human arm. It is used for converting the operator's arm movements into corresponding robot motion in real-time. To achieve this, the section proposes a model based on the principles of human kinematics. It integrates the motion data obtained in earlier chapters, along with anthropometric characteristics of the human body, to construct a comprehensive human motion control model. This model aims to estimate the position vector of the terminal joint of the arm. Furthermore, the model considers various factors that influence human motion, including joint constraints, the range of motion, and the dynamics of the human arm. By taking these factors into account, the tracking algorithm can more effectively capture the nuanced movements of the operator's arm, even in complex or highly dynamic teleoperation scenarios. In addition to providing precise motion tracking, the algorithm is designed to adapt to different human operators and varying ergonomic conditions. By utilizing personalized anthropometric data, the model can be customized to the specific physical characteristics of each user, ensuring that the motion control system is optimized for individual needs and capabilities. Ultimately, the goal of this approach is to enhance the efficiency and intuitiveness of human–robot interaction, and robot's movements are a seamless extension of the operator's own actions.

On the basis of the constructed human skeleton model, the local coordinate system of each limb segment is determined. The vector of each limb segment is vectorized, which represents the corresponding limb segment using a vector pointing from the proximal joint to the distal joint. It is represented as the child node limb segment vector in the local coordinate system of the parent node limb segment. The specific vector representation is shown in Fig. 3.1. The length of each limb segment vector is calculated according to "Human dimensions of Chinese adults" (GB/T 10000-2023).

The motion trajectory of the human arm's terminal joints needs to be represented in the global body coordinate system. It converts the limb segment vector l_b from the local to the global coordinate system, thereby obtaining the target limb segment's position and posture. The motion data q_s of the inertial sensing node is first resolved to obtain the real-time attitude quaternion of each inertial sensing node in the geographical coordinate system. Subsequently, the attitude transformation matrix obtained by attitude calibration is applied to determine the real-time posture quaternion q_b of each limb segment in the body coordinate system, which is finally combined with the real-time posture quaternion to convert the limb segment vector to the global coordinate system. The calculation method is as follows,

$$\left[0\ l_{g-i}^t\right] = q_{b-i}^t \otimes \left[0\ l_{b-i}\right] \tag{3.1}$$

where l_{g-i}^t is the real-time limb segment vector of the i_{th} limb segment in the global coordinate system at time t; q_{b-i}^t is the real-time pose quaternion of the i_{th} limb segment in the global coordinate system at time t; l_{b-i} is the vector representation of the i_{th} limb segment in the local coordinate system of the parent node; and $\otimes$ is the quaternion multiplication algorithm. l_{b-i} and l_{g-i}^t are represented in pure quaternion form. According to the kinematic solution sequence, the representation of each limb segment in the global coordinate system of the human body can be obtained through iteration,

$$p_t^m = \sum_{i=0}^{m} l_{g-i}^t \tag{3.2}$$

where p_t^m is the real-time position vector of the m_{th} joint in the human coordinate system at time t.

3.3 Human–Robot Motion Mapping and Planning

3.3.1 Human–Robot Workspace Mapping

In a master–slave heterogeneous system involving the human arm and the YuMi robot arm, the substantial differences in structure and joint degrees of freedom (DOF) present a complex challenge for designing an effective pose mapping strategy. The human arm is a highly sophisticated and flexible biomechanical system with multiple degrees of freedom across its joints. In particular, the shoulder and wrist joints enable compound motion. This rich range of motion makes the human arm an ideal model for teleoperation systems, where intuitive control and flexibility are

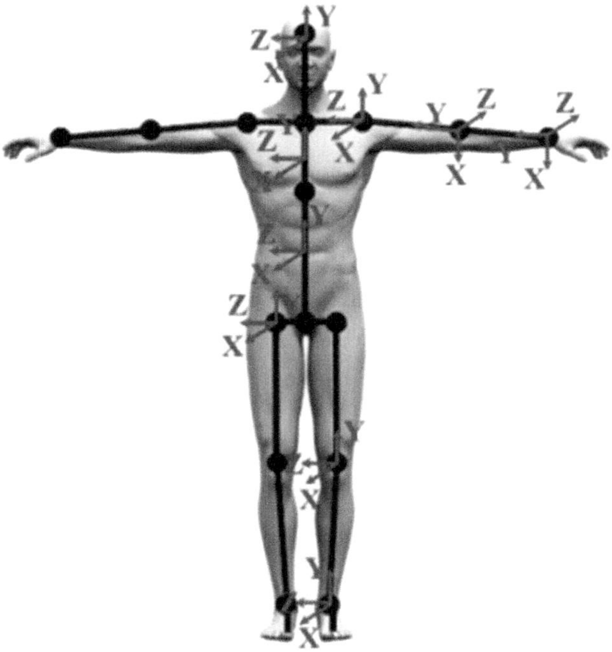

(a) Simplified skeleton model of human body

Limb segments	Limb segment vector	Limb segments	Limb segment vector
Thoracic vertebra	$[0 \quad l_{spine} \quad 0]$	Right foot	$[0 \quad -l_{r-foot} \quad 0]$
Lumbar vertebra	$[0 \quad l_{spine} \quad 0]$	Left shoulder	$[0 \quad 0 \quad -l_{l-shoulder}]$
Right shoulder	$[0 \quad 0 \quad l_{r-shoulder}]$	Left upper arm	$[0 \quad -l_{l-uparm} \quad 0]$
Right upper arm	$[0 \quad -l_{r-uparm} \quad 0]$	Left forearm	$[0 \quad -l_{l-lowarm} \quad 0]$
Right forearm	$[0 \quad -l_{r-lowarm} \quad 0]$	Left hip	$[0 \quad 0 \quad -l_{l-hip}]$
Right hip	$[0 \quad 0 \quad l_{r-hip}]$	Left thigh	$[0 \quad -l_{l-thigh} \quad 0]$
Right thigh	$[0 \quad -l_{r-thigh} \quad 0]$	Left calf	$[0 \quad -l_{l-shank} \quad 0]$
Right calf	$[0 \quad -l_{r-shank} \quad 0]$	Left Foot	$[0 \quad -l_{l-foot} \quad 0]$

(b) Partial limb segment vectors of the human body in an upright state

Fig. 3.1 Representing main limb segments based on the human coordinate system

paramount. In contrast, the YuMi robot arm, while designed with 7 DOFs to simulate the human arm's motion, faces limitations due to its mechanical structure and engineering design. These limitations include restricted joint motion ranges, reduced force control, and less adaptability in the end-effector's interaction with the environment. For example, while the human arm can move with great dexterity and flexibility, the YuMi arm's end-effector movements are constrained by its mechanical joints and lack the same natural fluidity and range as the human arm. Additionally, the robot arm may struggle to replicate the fine control needed for delicate operations that involve multiple joint articulations. This study proposes a novel pose mapping strategy that calculates the pose vector of the human arm's terminal joint and maps it to the robotic end-effector. This mapping process requires a careful balance between the kinematic and dynamic characteristics of both systems, ensuring that the robot arm's movements can closely align with those of the human operator, despite their structural differences. The pose vector of the human arm's terminal joint serves as a key reference point, translating the operator's arm motion into the robot arm's control space. The challenge lies in ensuring that this mapping not only preserves the spatial alignment of the human and robot arms but also accommodates the mechanical constraints and operational range of the YuMi arm. To achieve a more natural and accurate human–robot collaboration, the mapping strategy must incorporate real-time adaptability to dynamically adjust to changes in the operator's movements. For instance, in complex or rapidly changing tasks, the robot arm should be able to continuously adjust its posture and end-effector position to maintain synchronization with the human operator. This involves continuously analyzing and adjusting the robot's motion trajectory based on real-time data, while also factoring in the kinematic limits and torque capabilities of the robot arm. The goal is to ensure that the robot arm moves fluidly and accurately in response to the operator's inputs, without exceeding its mechanical limits or causing undesired jerks or misalignments in the motion. Furthermore, the pose mapping strategy must also consider the specific characteristics of the task being performed. For example, tasks requiring high precision, such as assembling small components or manipulating delicate objects, demand that the robot arm exhibits fine control and dexterity. In these cases, the mapping strategy should prioritize the replication of the operator's fine movements, adjusting the robot's motion to simulate the control required by the human fingers and wrists. In conclusion, a sophisticated pose mapping strategy is required in achieving a high degree of naturalness and accuracy in human–robot collaboration with a master–slave heterogeneous system. This strategy must address the inherent structural differences between the two systems, consider the dynamic and kinematic constraints of both the human arm and the robot arm, and ensure real-time adaptability to execute complex tasks effectively.

This section constructs a simplified human skeleton model and a robot arm model based on the structural characteristics of the operator's arm and robot arm (Fig. 3.2). Based on inertial motion data and kinematic calculations, the pose vectors of the end joints of the left and right arms of the human body in the global coordinate system were obtained. The real-time pose vector of the robot's end effector can be directly read from the controller. The workspace of a single human arm is approximately a sphere with a radius of r_h, and the workspace of a single robot arm is approximately a sphere with a radius of r_r, in order to obtain a ratio coefficient of k_{rh} between them,

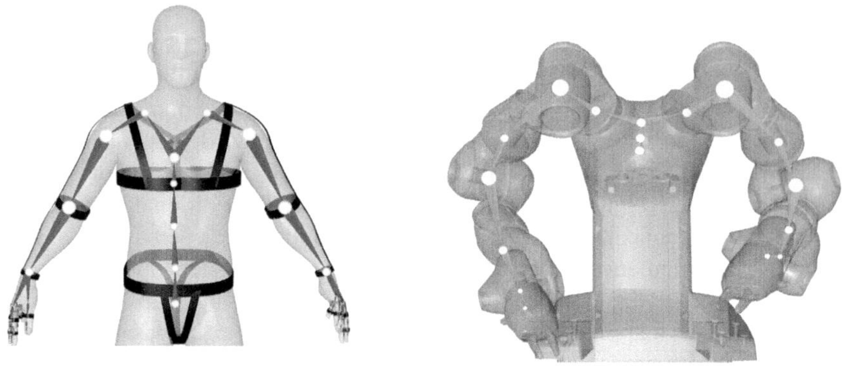

(a) Simplified skeleton model of human arms (b) Simplified model of robot arms

Fig. 3.2 Simplified models of the human arms and robot arms

$$k_{rh} = \frac{r_r}{r_h} \tag{3.3}$$

This section considers the movement of the human arm and the robot arm as rotations around their respective shoulder joints. Therefore, the motion of the human arm's end joint and the motion of the robot arm's end effector can be transformed into a local coordinate system centered at the shoulder joint. They are simplified as the human arm skeleton model and the robot arm model. Using the ratio coefficient k_{rh} derived from the workspaces of the human arm and the robot arm, the position representation of the human arm's end joint in the local coordinate system of the shoulder joint is transformed. This adjusted position is then combined with the known position vector of the robot's shoulder joint to determine the end effector's position vector in the robot's coordinate system. Simultaneously, the posture of the human arm's end joints is assumed equivalent to that of the robot arm's end effector, allowing for direct attitude mapping. The transformation relationship is detailed as follows:

$$k_{rh} \times \left(\begin{bmatrix} X_{Hh} \\ Y_{Hh} \\ Z_{Hh} \end{bmatrix} - \begin{bmatrix} X_{Hs} \\ Y_{Hs} \\ Z_{Hs} \end{bmatrix} \right) + \begin{bmatrix} Y_{Rs} \\ Z_{Rs} \\ X_{Rs} \end{bmatrix} = \begin{bmatrix} Y_{Rg} \\ Z_{Rg} \\ X_{Rg} \end{bmatrix} \tag{3.4}$$

$$q_{Hh} = q_{Rg} \tag{3.5}$$

where $[X_{Rg}, Y_{Rg}, Z_{Rg}]$, q_{Rg} are the position vector and attitude quaternion of the end effector of the robot arm after mapping adjustment; $[X_{Rs}, Y_{Rs}, Z_{Rs}]$ is the position vector of the robot's shoulder joint in the global coordinate system of the robot; $[X_{Hh}, Y_{Hh}, Z_{Hh}]$, q_{Hh} are the position vector and posture quaternion of the human arm end joint in the shoulder coordinate system; $[X_{Hs}, Y_{Hs}, Z_{Hs}]$ is the position vector of the human shoulder joint in the shoulder coordinate system. Based on the above transformation relationship, we need to transform the pose vector of the human arm end joint calculated in the previous chapter to obtain the pose vector of the robot end effector.

3.3.2 *Motion Optimization and Planning*

In teleoperation systems, the primary target is to ensure that the robot's movements not only accurately reflect the operator's intentions but also conform to the robot's inherent kinematic and dynamic capabilities. Achieving this requires a sophisticated pose mapping strategy that translates the operator's motions, particularly from the human arm's terminal joint, into corresponding robotic actions. It also needs to maintain the naturalness and fluidity of the human motion. The foundation of this approach lies in a well-defined pose mapping relationship between the human and robot workspaces. By directly mapping the pose vector of the human arm's terminal joint to the robot's end-effector, we ensure that the robotic movements closely align with the operator's physical motion. The robot's control system must be designed to interpret these mappings and execute them with the constraints of the robot's own mechanical and computational framework. To further enhance the anthropomorphism of the robot's motion trajectory, we introduce motion constraints specific to the robot's capabilities. These constraints are derived from the robot's kinematic properties, for example, workspace, and ensure that the robot's movement is feasible and safe within its design limitations. For instance, while the human arm is highly flexible and capable of executing a wide range of precise movements, the robot arm must operate within its predefined workspace and joint limits, which may not offer the same level of flexibility. Therefore, the motion planning strategy incorporates these constraints to prevent the robot from attempting motions that exceed its physical limits or lead to jerky, unnatural movements. The motion planning strategy aims to strike a balance between the operator's input and the robot's mechanical limitations, ensuring smooth, coordinated movements. This involves using advanced algorithms that predict and adapt to the operator's movements in real-time, adjusting the robot's trajectory accordingly. The planning strategy also incorporates feedback mechanisms to continuously refine the robot's movements, and it takes real-time sensor data, environmental changes, and task-specific requirements into account. By incorporating these motion constraints and planning strategies, teleoperation systems can significantly improve both the accuracy and efficiency of robot movements, ensuring that the robot's actions are both faithful to the operator's intentions and compatible with its mechanical capabilities. As teleoperation systems become more widely deployed across industries such as smart manufacturing, healthcare, and hazardous environments, the ability to seamlessly translate human motions into robotic actions, while ensuring the robot's movements are safe, efficient, and aligned with human capabilities.

In this study, the pose mapping between the human arm and the robot arm, focusing solely on the terminal joint, often results in a significant discrepancy between the robot arm's motion trajectory and the human arm's actual trajectory. The specific reasons are as follows.

(1) Deficiency in environmental perception: The robot arm manifests a prominent limitation concerning its capacity to sense the surrounding environment. It conspicuously lacks the requisite faculties to precisely apprehend the external environment. Absent state-of-the-art sensors and intricate perception algorithms, it remains incapacitated in detecting and reacting to obstacles that impede its operational trajectory. Consequently, during task execution, its inability to autonomously and efficaciously execute evasive maneuvers or formulate reliable prognoses regarding potential impediments leads to an inevitable breakdown in the motion planning phase, which is elemental for seamless robot arm functionality. In teleoperated scenarios, where the human operator's purview of the robot arm's actual operating environment is often circumscribed, the environmental complexity exacerbates this predicament. The robot arm is obligated to exhibit a degree of autonomous intelligence, necessitating it to possess an aptitude for intelligent perception.
(2) Limitations of pose mapping strategy: In the domain of robotic control and motion planning, the existing pose mapping strategy exhibits several notable constraints. Currently, these approach in pose mapping predominantly focus on the determination and pursuit of the desired pose of the end-effector. However, it engenders a critical oversight as it blatantly disregards the intricate kinematic characteristics inherent to the intermediate joints of the robot arm. The kinematics of these intermediate links contribute to the propagation and modulation of forces and velocities. Moreover, this prevalent strategy fails to take into account the kinematic features embedded within human motion patterns. Humans, in their natural movements, exhibit a seamless coordination and adaptation among multiple joints, which is underpinned by complex kinematic relationships. By neglecting these anthropomorphic kinematic cues, the robot arm, operating under the current pose mapping regime, is prone to generating path points that deviate significantly from the optimal and natural trajectories. This deficiency becomes particularly pronounced in applications where the robot arm is required to mimic human-like actions or operate in environments that demand dexterous and smooth maneuvers, such as in surgical robotics or delicate assembly tasks.

Humans plan the motion paths of the left and right arms according to the motor coordination system to avoid mutual collision and interference, coordinating the arms to complete operations together. Accordingly, this section proposes the motion planning strategy of the dual-arm robot based on the characteristics of human arm motion. It not only considers the pose of the end-effector but also imposes constraints on the motion of the intermediate joints to conform to the kinematic characteristics of the human body. By imitating the motion constraints of the human arm and using the method of constructing virtual obstacles, workspace constraints are added to each robot arm to ensure that the motion trajectory of the robot arm is more in line with human physiological characteristics.

This study takes ABB's first collaborative robot YuMi as an example, establishing motion constraints based on its motion planning trajectory. When the constraints of collaborative motion for the two arms are not added, the joints of the two mechanical arms may collide, leading to the failure of the motion planning for the two arms. Therefore, a virtual barrier is constructed between the two arms by imitating the characteristic of human arms moving without interfering with each other. The redundant degrees of freedom of the dual-arm robot and the larger workspace of the mechanical arm compared to the workspace of the human arm may result in the path planning of the mechanical arm not conforming to the optimal control principle of human motion, such as complex planning paths for rotational movements. The mechanical arm performs repetitive rotational movements around the shoulder joint. Therefore, a virtual barrier is set above the mechanical arm to constrain the workspace of the two arms. The motion planning of the mechanical arm does not meet the requirements of the shortest path. When the robot is close to the target object, it still chooses a complex path. By setting virtual obstacles on both sides of the mechanical arm, unnecessary motion planning is reduced. The study found that in human–robot teleoperation interaction, the anthropomorphism of mechanical arm motion planning can be effectively improved by drawing on the characteristics of human somatic motion.

3.4 Experiment and Results

(1) Human–robot workspace mapping transformation

In human–machine kinematic analysis, establishing the forward kinematic models of the human body and the robot is important. In this experiment, the Denavit-Hartenberg (D-H) method can be used to study a single human arm. It establishes the coordinate system of the human arm links, and calculates the pose transformation of the terminal joint of the human arm relative to the shoulder joint, as shown in Fig. 3.3a. Table 3.1 is the established D-H parameter table. Similarly, taking a single arm of a dual-arm robot as the research object, the coordinate system of the robot links is established, the relationships between each link are established, and the pose transformation of the robot's end-effector relative to the base is derived. The link coordinate system constructed using the D-H parameter method is shown in Fig. 3.3b, and according to the established link coordinate system, the parameters of each link are determined, and the established D-H parameter table is shown in Table 3.2.

The Monte Carlo method, renowned for its efficacy in handling complex stochastic problems, has been extensively employed in the domain of robotics to ascertain the workspace of robotic systems. In this context, the approach initiates by randomly generating copious joint angle combinations. These combinations are not arbitrary in a haphazard sense; rather, they span a wide spectrum that is theoretically representative of all possible articulations of the robot's joints. By using the forward

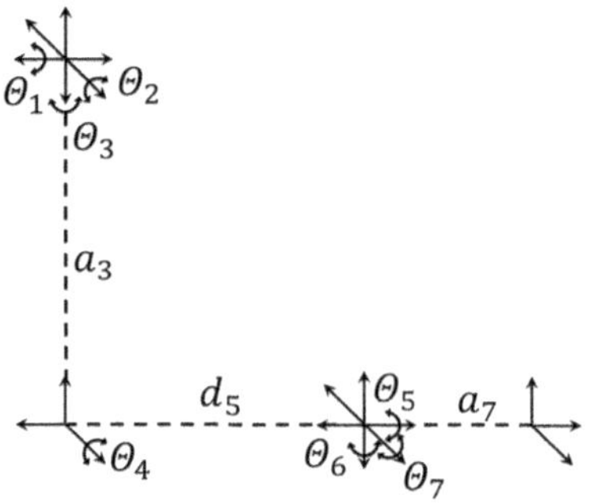

(a) Coordinate transformation of human arm D-H model

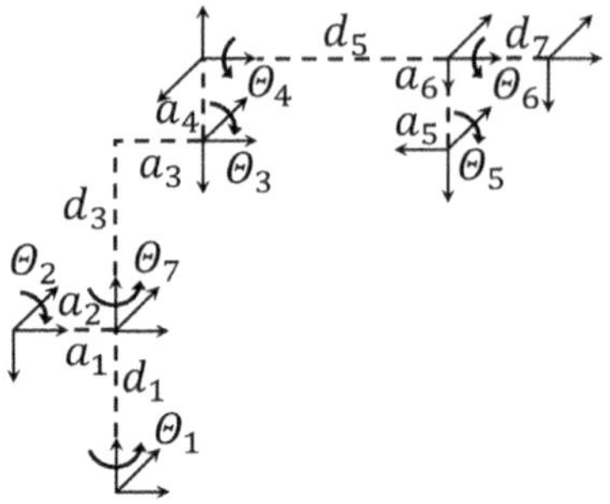

(b) Coordinate transformation of robot arm D-H model

Fig. 3.3 D-H models of human arm and robot arm

Table 3.1 D-H parameters of human arm

i	a_{i-1}/mm	α_{i-1}/°	d_i/mm	θ_i/°	Joint angle range/°
1	− 30	− 90	166	0	± 168.5
2	30	90	0	0	− 143.5 ~ 43.5
3	40.5	− 90	251.5	0	− 123.5 ~ 80
4	40.5	− 90	0	− 90	± 290
5	27	− 90	265	180	− 88 ~ 138
6	− 27	90	0	0	± 229
7	0	0	36	0	± 168.5

Table 3.2 D-H parameters of robot arm

i	a_{i-1}/mm	α_{i-1}/°	d_i/mm	θ_i/°	Joint angle range/°
1	0	− 90	0	− 90	± 90
2	0	− 90	0	− 90	− 180 ~ 50
3	260	0	0	90	− 180 ~ 80
4	0	90	0	90	− 10 ~ 145
5	0	90	280	90	± 90
6	0	90	0	90	− 90 ~ 70
7	100	0	0	0	− 15 ~ 40

kinematic model, the position and posture of the end-effector corresponding to each of these randomly generated joint angle sets can be precisely calculated. The forward kinematic model encapsulates the geometric and kinematic relationships between the robot's joints and its end-effector. It transforms the angular displacements at the joints into the Cartesian coordinates and orientation matrices that define the location and pose of the end-effector in three-dimensional space. Through a repetitive application of this model to the numerous joint angle combinations furnished by the Monte Carlo sampling process, a comprehensive dataset of end-effector positions and postures is acquired. The workspace is defined as the spatial expanse that the robot end-effector can feasibly reach under the purview of all conceivable joint angle permutations. It is a volumetric region that delineates the robot's operational envelope, dictating the scope of tasks it can undertake. Employing the Monte Carlo method, we are empowered to computationally simulate these random joint angle combinations within a virtual environment, typically on a computer. This simulation enables us to observe and record the distribution of the end-effector across the three-dimensional space. Subsequently, by collating and processing this spatial distribution data, a graphical representation of the workspace can be meticulously crafted. This graphical depiction, often in the form of a three-dimensional plot or a series of two-dimensional projections, provides an intuitive visualization of the robot's reachability.

In this experiment, the MATLAB Robotics Toolbox was utilized to build the kinematic model. For analyzing the robot's workspace, the Monte Carlo method was adopted. When dealing with multiple joints and links, both human and robot kinematic models are complex. The Monte Carlo method bypasses solving complex equations by random sampling. It generates numerous random joint angle combinations, feeds them into the forward kinematic model to get end-effector positions and postures, and through aggregation, approximates the workspace.

Given that each kinematic model parameter has a corresponding range of values, the Monte Carlo method traverses the values of each joint within their respective range, generating random values at the endpoints to form the workspace.

The specific operational method involves solving the kinematics positive solution expressions of the robot arm and the human arm based on the kinematic model. Furthermore, random values are generated within the numerical range of each joint variable. The above random variables are applied to the forward solution equation to determine the position vector of the endpoint. The MATLAB's visualization tool is used to draw the corresponding workspace in which the value of each joint variable is generated by sampling pseudo-random numbers [1],

$$\theta_i = \theta_{imin} + (\theta_{imax} - \theta_{imin}) \times rand(n) \tag{3.6}$$

where $i \in [1, 2, 3, \ldots, 7]$, $n \in N^*$, θ_i is the numerical value obtained by Monte Carlo method for the i_{th} joint, θ_{imin} and θ_{imax} are the minimum and maximum values corresponding to the i_{th} joint, $rand(n)$ is a pseudo-random number that follows continuous and uniform distribution.

Figure 3.4 shows the workspace of the human arm and robot arm obtained by using numerical methods. It is the original workspace of the human arm and robot arm, with part of the workspace of the human arm not covered by the robot arm workspace. According to the above formula, obtain the ratio coefficient of two workspaces, $k_{rh} = 1.21$.

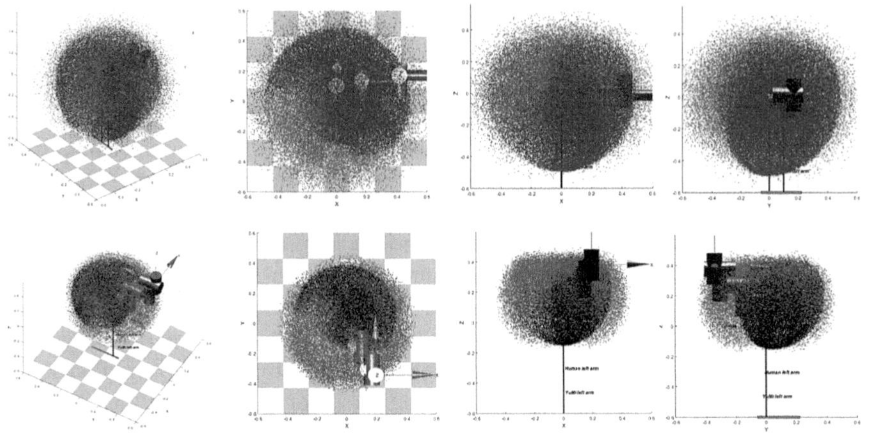

●Robot origin workspace ●Human origin workspace ●Adjusted robot workspace

Fig. 3.4 Three-dimensional workspace of human arm and robot arm

Using the human motion mapping strategy, the position vector of the human arm's end joint is transformed and mapped to the robot's end using the proportion coefficient k_{rh}, and the transformed robot workspace is obtained, as shown in Fig. 3.4. After conducting a comparative analysis between the workspace of the human arm and that of the robot arm, both in its original state and subsequent to the implementation of transformation measures, it becomes evident that the application of a human motion mapping strategy serves as a potent means to efficaciously modulate the robot's workspace. This strategic intervention enables the robot arm's workspace to be tailored in such a manner that it closely approximates the kinematic capabilities and spatial reach of the human arm. Such an approach not only bridges the gap between the operational envelopes of the two systems but also paves the way for enhanced ergonomic compatibility and seamless interaction in human–robot collaborative scenarios. This is predicated on the premise that a more anthropomorphic workspace for the robot arm facilitates the emulation of natural human motion patterns.

(2) Humanoid motion path planning

In the realm of collaborative dual-arm robots, the motion planning strategy is designed to imitate the quintessential characteristics of human arm motion planning. This is achieved by integrating spatial constraints into the kinematic framework of the mechanical arms. The strategic emplacement of virtual obstacles emerges as a pivotal determinant in safeguarding the efficacy of path planning [2]. Particularly in teleoperation scenarios, despite the absence of tangible physical impediments, the anticipation of potential environmental hazards empowers us to embed obstacle avoidance logic within the robot's virtual motion planning paradigm. To rigorously analyze the influence of spatial constraints on path planning, this experimental investigation undertakes a series of comparative experiments. The objective is to analyze the relationship between the imposed spatial constraints and the resultant path planning outcomes. Through this approach, we endeavor to identify the most suitable spatial constraint scheme for the current study. This determination process is predicated on a comprehensive consideration of multiple factors, including operational efficiency, safety, and the naturalness of motion. We want to realize optimal motion planning for collaborative dual-arm robots engaged in teleoperation tasks, thereby enhancing their overall performance and compatibility in complex working environments.

Prior to the initiation of the experimental protocol, the participants wear the inertial motion capture device according to the standardized operating procedures. Subsequently, a comprehensive suite of calibration procedures for joint postures and kinematic states were executed. This aims to ensure the utmost precision and reliability of the ensuing data acquisition phase, minimizing potential errors and artefacts that could skew the results. The experimental robot was ABB's inaugural collaborative robot, the YuMi, renowned for its sophisticated redundant design architecture. Each of its manipulator arms is outfitted with 7 DOFs and a total of 14 DOFs for the bimanual configuration. This redundancy endows the robot with enhanced manipulability and adaptability, making it suitable for complex human–robot interaction scenarios in collaborative workspaces. During the experiment, the participants executed precise arm control maneuvers to effectuate the grabbing of the target object, conforming precisely to the predefined arm kinematic configurations established in the experiment. Concurrently, the inertial motion capture system established a robust communication link with the robot's control system via a local area network. This enabled the real-time streaming of the captured human motion data, replete with kinematic parameters to the robot's onboard processing unit. In the domain of robot motion planning, the experiment divided the path planning process into two discrete subsets corresponding to the left and right manipulator arms. In response to the overarching high-level motion directives, the robot was actuated to perform coordinated motion planning and to synchronously track the dynamic trajectories of the human arm. To further analyze the impact of spatial constraints on robot motion efficacy, the experiment incorporated three distinct classes of spatial constraints into the planning parameters. These constraints were designed to adjust the robot's workspace, taking into account factors such as obstacle avoidance, joint limits, and ergonomic considerations. By employing the metric of motion path length and motion planning time as key quantitative indicators, a comparative analysis was conducted to evaluate the relative merits of the different spatial constraint formulations. This methodological approach aimed to clarify the optimal spatial constraint strategy for enhancing the performance and efficiency of the robot in the given human–robot collaborative task.

The experimental task requires subjects to complete grabbing. The movement trajectory includes four points: the starting point P0, the intermediate waypoint P1 and P2, and the endpoint P3. The robot's movement trajectory mainly includes three parts, namely P0 to P1, P1 to P2, and P2 to P3, set the three path points as target points in the robot path planning, and superimpose the three path planning times. The experiment shows the movement trajectory of the robot under different obstacle settings (Fig. 3.5). The four obstacle settings are (1) no virtual obstacles, (2) large size obstacles, for example, 660 mm $\times$ 750 mm (width $\times$ height), (3) medium size

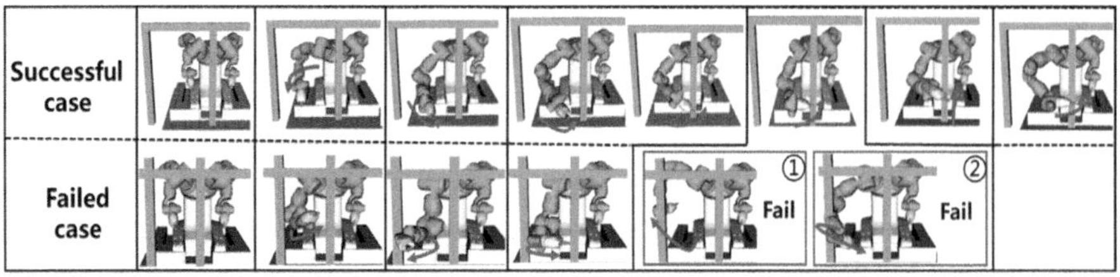

Fig. 3.5 Motion planning path for setting virtual obstacles of different sizes

obstacles, for example, 420 mm × 600 mm, (4) small size obstacles, for example, 300 mm × 500 mm. Utilizing small-size obstacle constraints results in a lower success rate for robot motion planning, particularly when obstacles are narrower than 300 mm and shorter than 600 mm, leading to frequent collisions or penetrations. This is because small-size obstacles have more constraints in the robot workspace, leading to a reduction of paths available for robot motion planning. When large-sized obstacles are used, the robot's achievable motion planning paths increase, and the success rate of motion planning increases. However, since there are no constraints on the motion of the middle joint of the robot arm, the humanoid nature of the robot arm's motion path is poor. In the absence of motion constraints, the robot motion planning path is similar to that of large-size constraints, and the robot arm motion path is less human-like. In contrast, medium-sized obstacles can better constrain the robot's movement path. It is similar to the movement trajectory of the operator's arm. Therefore, the results indicate that the representative case is "low human-like", while the experimental results using medium-size virtual obstacles is "high human-like".

Figure 3.6 illustrates the time required for motion planning with virtual obstacles of varying sizes, with widths ranging from 300 to 660 mm and heights from 500 to 750 mm. The experiment displays the variation curve of motion planning time under settings of virtual obstacles of different sizes. Analysis indicates that increasing the spatial constraint dimensions reduces the time for path planning. For instance, at a height of 700 mm, the planning time for an obstacle measuring 660 mm in width is significantly shorter than for one measuring 300 mm in width. For instance, when the height is set at 700 mm, the planning time for an obstacle with a width of 660 mm is significantly shorter than that for an obstacle with a width of 300 mm. Considering the anthropomorphic nature of the robot arm's motion trajectory and the real-time requirements of motion planning, the optimal spatial constraint dimensions identified in this experiment are 420 mm in width and 600 mm in height.

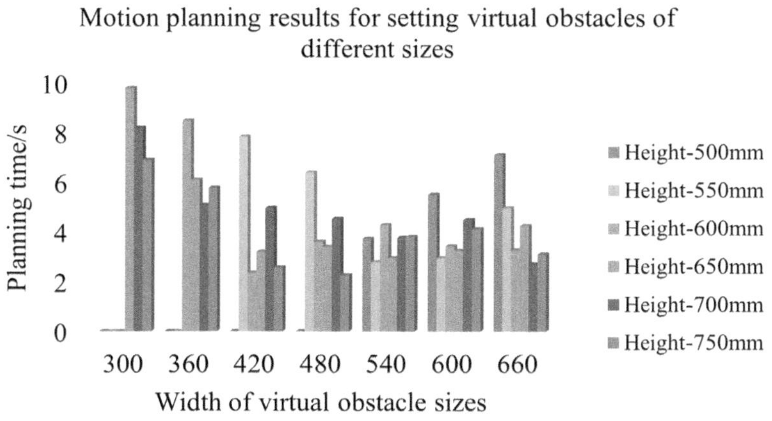

Fig. 3.6 Motion planning results for setting virtual obstacles of different sizes

3.5 Implementation

In the era of smart manufacturing, teleoperation technology integrates modern control theory, computer science, human–robot interaction, and robotics. This integration allows operators to remotely manipulate robots or other equipment through control interfaces, enabling human operators to overcome physical and environmental limitations. Teleoperation enables operators to control remote equipment with flexibility, making it possible to perform complex or delicate tasks that would otherwise be challenging. Moreover, in hazardous environments or locations inaccessible to humans directly, such as in nuclear accident or deep-sea exploration, teleoperation technology enhances safety by minimizing human presence and improving work efficiency. Similarly, in specialized scenarios of smart manufacturing, such as high-temperature, high-pressure industrial environments or radioactive production areas, teleoperation technology plays a crucial role in maintaining operational safety. The significance of research in remote operation technology lies in its potential to enhance safety, flexibility, and efficiency while broadening the scope of human activity applications. It contributes to elevating the smart manufacturing to a higher level.

As technologies such as 5G and artificial intelligence advance, the future of human–machine teleoperation in smart manufacturing will increasingly emphasize intelligence and precision. This focus will offer more efficient and accurate operational modes for manufacturing, providing new momentum into both production and broader societal development.

3.5.1 Human–Robot Teleoperation in Laboratory

In the laboratory setting, the hardware infrastructure for the human–machine teleoperation interactive application comprises two principal components: the human motion capture system and the robot system. The human motion capture system is an intricate assemblage of multiple elements. It incorporates inertial motion capture device, data gloves, wireless transmission modules, and associated software modules. The inertial motion capture device, using advanced inertial sensor technologies, is designed to track and record the motion of the human body. The data gloves are equipped with a suite of sensors sensitive to finger flexion and extension, and the human body movements and intricate hand gestures is acquired. These captured data, facilitated by the software modules that perform functions such as data filtering, noise reduction, and feature extraction, are then transmitted in real-time via the wireless transmission modules. The wireless transmission modules ensure reliable and high-speed data transfer, thereby enabling seamless interaction between the human operator and the robotic counterpart. The robot system centers around ABB's YuMi dual-arm robot, a robotic platform renowned for its high dexterity. Each individual robot arm is engineered with 7 DOFs, endowing it with the capacity to execute a wide array of complex motions. Additionally, the gripper attached to each arm is equipped with 1 DOF for grabbing and manipulating objects. The robot communicates with the main controller through industrial Ethernet. It is a communication medium renowned for its robustness and high bandwidth. The IRC5 controller, acting as the brain of the robotic system, is pre-installed with the RobotWare operating system. This operating system including algorithms for motion control, path planning, and collision avoidance, orchestrates the coordinated movements of the two arms of the YuMi robot.

The dual-arm design of the YuMi robot not only imitates the movements of the human hands, providing excellent coordination and parallel operation capabilities, but also integrates advanced control algorithms to achieve synchronization and collaborative work between the two mechanical arms. This enables YuMi to excel in performing complex tasks. In terms of design, YuMi allows for flexible deployment in spatially constrained working environments while maintaining a high-performance motion range and load capacity. Its high-precision motors and transmission systems ensure the accuracy of operations. Moreover, YuMi's software platform supports open programming and can be integrated into various industrial automation systems. It offers users a broad space for customization and expansion.

This application is developed based on the Robot Operating System (ROS), employing ROS-Industrial to build a control framework. Based on the Simple Message protocol, two robot arms and grippers are configured in the application layer of the controller to ensure communication and control. The official program package ABB-driver for ABB industrial robots for ROS communication is configured in the ROS interface layer. The motion control and planning of the robot are completed based on the ROS system and MoveIt. Since the control of the robot arm and gripper is based on different IP addresses, this application divides the robot into four motion planning groups, i.e., left arm, right arm, left hand, and right hand. According to the robot's starting state and target state, the inverse kinematics solution of the robot is obtained and motion planning is completed. In the teleoperation interactive application in the laboratory, human–machine motion mapping first calculates the human motion trajectory from inertial motion data. According to the human motion mapping strategy, the end joint position of the human arm is mapped to the robot end. The motion planning takes the current position of the robot as the starting state, and the transformed position of the end joint of the human arm is taken as the target state. The robot joint value is solved using a motion solver.

In this particular section, the utilization of the aforementioned system is demonstrated for the implementation of a remote grabbing application. Both the robot and the human operator are situated within the laboratory environment. The operator, equipped with an inertial motion capture device, undertakes the task of grabbing target. The motion data, which has undergone comprehensive processing procedures to enhance its accuracy and reliability, is then transmitted to the robot. Subsequently, the robot arm is designed to mimic and follow the movements of the human arm, thereby executing the object-grabbing task. As illustrated in Fig. 3.7, the operational task can be dissected into four primary and sequential steps: "preparation", "approach", "grip", and "pick". Initially, upon the operator reaching the ready state at the starting position, they control their arms to initiate the approach towards the target object. Concurrently, the robot, in synchronization with the operator's actions, maneuvers to the requisite operational position. During the operation, the operator's hand and the robot's gripper maintain a coordinated state of being either open or closed, depending on the specific requirements of the task. Once the target position is arrived, the robot's gripper is precisely controlled to open and close, facilitating the successful grabbing of the target object. Upon the right-side operation, the operator proceeds to control the left arm, repeating the aforementioned process in a similar fashion to finalize the overall operation task. This setup not only showcases the practical application of the integrated system but also highlights the seamless coordination required between human and robotic entities to achieve efficient object manipulation in a remote-controlled context.

3.5.2 Human–Robot Intercontinental Teleoperation

This section employs a China-Sweden intercontinental teleoperation to validate the remote interactive application. The system includes a human operator located in Västerås, Sweden and a robotic system located in Hangzhou, China (Fig. 3.8). This example uses a self-developed inertial motion capture device to obtain human body motion and uses human motion data to remotely operate a robot to perform operational tasks. The human motion capture system includes inertial motion capture equipment, feedback devices, wireless communication modules, and supporting software. Wearable devices include feedback devices, such as vibration modules and light-emitting diode modules, to feedback on the operating status of the remote

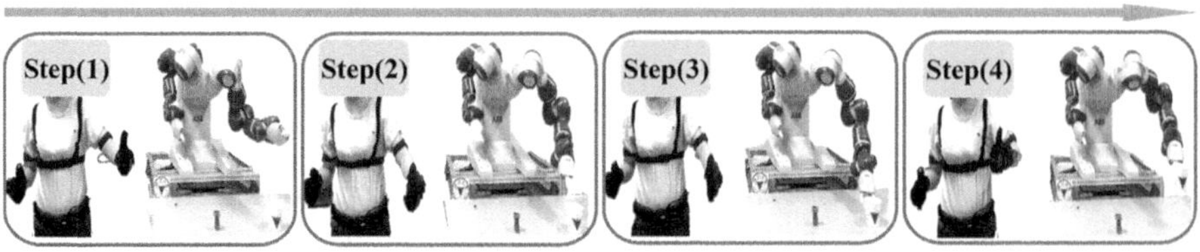

(a) Operator teleoperates robot to grab cup

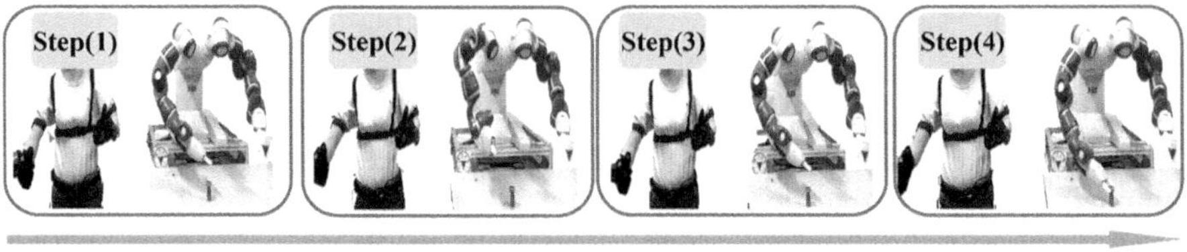

(b) Operator teleoperates robot to grab bottle

Fig. 3.7 Operator wearing inertial motion capture system remotely operates robot to grab target objects

robot and the perceived approaching status of external obstacles to the operator. A self-designed and integrated robot system is used in remote interaction applications. The robot includes two 6-DOF Kinova Jaco2 robot arms. The maximum angular velocity of the first three joints is 36°/s, and the maximum angular velocity of the last three joints is 48°/s. The arms are placed face-to-face in a "V" shape, imitating the configuration of human arms. Each arm has a maximum continuous load of 1.6 kg, a maximum workspace of 900 mm, a maximum clamping force of 40N, and a power consumption of 25 W. Each arm joint has a torque sensor for robot arm force and torque measurement. The robot's head is equipped with a stable camera, positioned midway between two robot arms, with image acquisition and video streaming capabilities. The high-definition image transmission system has achieved ultra-low latency video transmission (10–40 ms) and remote data transmission (2–5 km). The robot is installed on an omnidirectional mobile platform with built-in laser radar, which can realize positioning, navigation, obstacle avoidance and other functions. Wearable devices and robots are connected to their respective local area networks through wired or wireless methods. Two-way communication is completed with the help of cloud servers. ROS is used as a platform for communication, control, and development. Wearable devices are connected to the ROS platform in the form of nodes to achieve information and service communication with the robots. In remote teleoperation applications, the operator's movements, collected and preprocessed through an independently developed inertial motion capture system, are transmitted to the robot, subsequently controlling the robot arm's movement based on human–machine motion mapping and planning strategies.

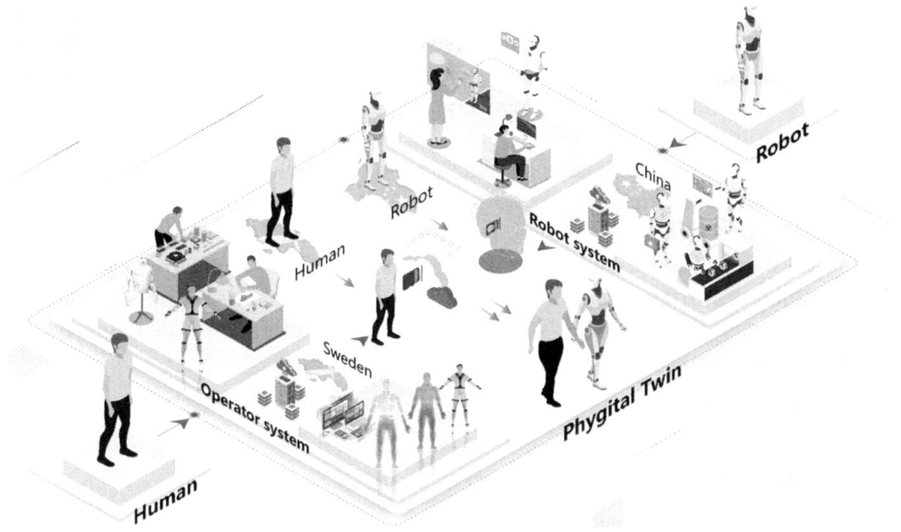

Fig. 3.8 Framework of human–robot intercontinental teleoperation

In this section, the above-described system is presented in a China-Swedish cross-continental teleoperation interactive application. The operator is geographically positioned in Västerås, Sweden, while the robot system is deployed in Hangzhou, China. Using custom-developed wearable devices, operators are empowered to remotely commandeer robots and carry out target tasks, thus facilitating cross-regional production collaboration. The status of the robot's end effector, along with the perceived approaching status of external obstacles, is relayed back to the operator through feedback mechanisms embedded within the wearable device. Simultaneously, the camera installed on the robot serves as visual feedback, providing real-time insights into the operation site. As depicted in Fig. 3.9, teleoperation tasks are exemplified. In the first task, the operator directs the robot to perform a task of grabbing and placing fruit. The active robot arm assumes the primary responsibility of grabbing the fruit, while the passive robot arm coordinates by moving the container box to accommodate the fruit. This coordinated effort requires precise motion control and synchronization between the two arms to ensure the efficient completion of the task. Feedback mechanism is primary in the human–robot interaction. It provides information for controlling strategy and giving warning message for protecting robots. Basic feedback strategy attempts to make immersive operation experience for operators. In the validation tasks, basic feedback accessories are equipped with operators. For example, when operators control the transnational robot, robotic perception information is transmitted to operators. Specifically, operator need to know if the robotic grippers open or not according to results of grabbing tasks. With the acquired information from vibration and flash, direct feedback information helps operator make decision conveniently. Feedback mechanism not only faces to operators, but also serves interactors. Robot skin aims to guide interaction existing in the safe zone. In the validation case, robot is equipped with robotic skin and user is equipped with shining led. When unknown objects are close to robot avatar, intelligent skin presents the potential danger for avoiding accident (Fig. 3.9). Overall, this cross-continental teleoperation application showcases the potential of integrating advanced technologies to overcome geographical barriers and achieve efficient human–robot collaboration in complex manufacturing scenarios.

The experiments effectively demonstrate the integration of physical and digital domains, establishing a foundation for a new generation of human–robot interaction in smart manufacturing. Smart manufacturing emphasizes the seamless merging of the physical and digital worlds, with these results providing innovative approaches to bolster human–machine collaboration.

3.6 Conclusion

This chapter focus on the capture and reproduction of human movement and its integration with smart manufacturing concepts. By developing advanced human motion mapping interaction technology, this research opens new possibilities for remote control and operation within smart manufacturing systems. Utilizing human kinematics models and inertial motion capture technology, we have achieved precise tracking of target joint motion trajectories. This advancement is vital for smart manufacturing applications, as it enables machines and robots to more accurately simulate human operator actions. Additionally, the human–robot skeleton model established provides a robust theoretical foundation for attitude mapping, facilitating highly synchronized teleoperation in manufacturing environments. To further enhance the naturalness, efficiency, and safety of teleoperation, we have designed an anthropomorphic motion mapping strategy. Experimental results indicate that the technology performs exceptionally well not only in laboratory environments but also demonstrates effectiveness and adaptability in intercontinental scenarios. In summary, this chapter provides both theoretical and practical support for teleoperation technology in smart manufacturing. By accurately capturing and reproducing human motion and implementing anthropomorphic motion mapping strategies, we are paving the way for a more efficient, safe, and natural future in smart manufacturing.

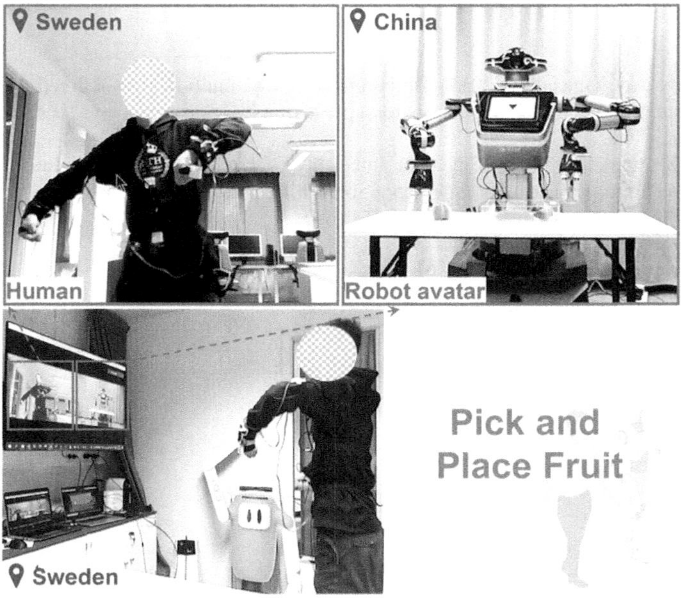

(a) Remote operation of wooden block grabbing

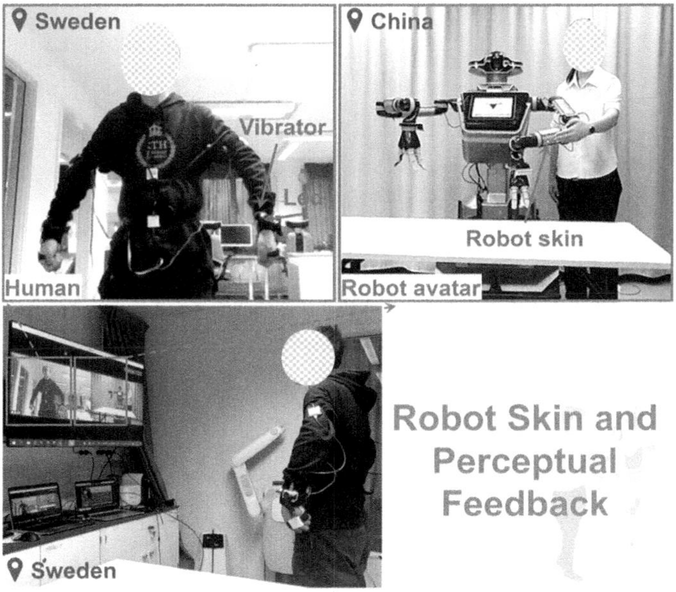

(b) Safety and feedback parts of the interactive system

Fig. 3.9 Operators wearing self-developed motion capture device in Sweden remotely operate robots in China

References

1. Chaudhury AN, Ghosal A (2017) Optimum design of multi-degree-of freedom closed-loop mechanisms and parallel manipulators for a pre scribed workspace using Monte Carlo method. Mech Mach Theory 118:115–138
2. Sharifi M, Chen X, Pretty C, Clucas D, Lunel EC (2018) Modelling and simulation of a non-holonomic omnidirectional mobile robot for offline programming and system performance analysis. Simul Model Pract Theory 87:155–169

Chapter 4
Human–Robot Collaboration Based on Motion Estimation

Abstract In line with a human-centric smart manufacturing vision, human–robot collaboration is striving to combine robots' high efficiency and quality with humans' rapid adaptability and high flexibility. In particular, perception, recognition and estimation of human motion determine when and what robot to collaborate with humans. This work presents an attention-based deep learning approach for inertial motion recognition and estimation in order to infer when robotic assistance will be requested by the human and to allow the robot to perform partial human tasks. First, in the stage of motion perception and recognition, quaternion-based calibration and forward kinematic analysis methods enable the reconstruction of human motion based on data streaming from an inertial motion capture device. Then, in the stage of motion estimation, residual module and Bidirectional Long Short-Term Memory module are integrated with proposed attention mechanism for estimating arm motion trajectories further. Experimental results show the effectiveness of the proposed approach in achieving better recognition and estimation in comparison with traditional approaches and existing deep learning approaches. It is experimentally verified in a laboratory environment involving a collaborative robot employed in a small part assembly task.

Keywords Human–robot collaboration · Inertial motion capture · Attention-based deep learning · Motion trajectory

Revised from Zhou et al. An Attention-based Deep Learning Approach for Inertial Motion Recognition and Estimation in Human–robot Collaboration, Journal of Manufacturing Systems, Copyright (2023), with permission from Elsevier Ltd.

Author: Huiying Zhou, Geng Yang, Baicun Wang*, Xingyu Li, Ruohan Wang, Xiaoyan Huang, Haiteng Wu, and Xi Vincent Wang.

H. Zhou et al., *Revitalizing Human-Robot Interaction*, Advanced Topics in Science and Technology in China 3, https://doi.org/10.1007/978-981-96-6139-8_4

4.1 Introduction

Nowadays, the need to comply with the shorter product life cycle, diversified market demands and growing global competitiveness lead to increasing requirements on manufacturing systems in terms of flexibility and diversity. Meanwhile, manufacturing systems in the context of Human-Cyber-Physical Systems (HCPSs) are steeping towards the vision of Human–Robot Collaboration (HRC), which facilitates flexible production by combining robots' high efficiency and quality with humans' rapid adaptability and high flexibility [1, 2].

Many existing assembly tasks pay much workload on humans using manual operation completely. With the involvement of robots, it is challenging for them to tackle the whole assembly task independently although robot is endowed with high efficiency and quality. Human–Robot Collaborative Assembly (HRCA) meets the requirements of efficiency of operations, resource utilization and intensity of work against aforementioned issues [3]. Thereinto, perception and estimation of human motion play a vital role in HRCA, which triggers robots to determine "when to deliver which targeted object to where" and execute motion plans for aiding operators. Further exploration and improvements on motion recognition and estimation for HRCA have been reported recently, such as from operator's motion classification to operator's motion tracking, from the non-learning-based robot control methodologies to model-based reinforcement learning variable impedance control methodologies, and etc. [4, 5]. Obviously, in terms of motion capture device, visual sensors-based device becomes the popular option for its accuracy and convenience, whereas it is accompanied by high cost, narrow application space, and risk of privacy disclosure. Wearable device acquires increasing attention for its convenience and applicability [6]. Nevertheless, major efforts have been put to handle classification of activities using wearable sensors without much consideration for motion trajectory cognition in industrial applications. Hence, an imperative research issue is how to develop an effectual motion recognition and estimation strategy using wearable device to facilitate robot to comprehend operation behavior.

This work hopes to contribute to human motion perception, recognition and estimation for HRCA by providing a new perspective. It achieves cognition of human behavior by integrating a hybrid approach based on forward kinematics and an attention-based deep learning model, with a developed inertial motion capture system. Meanwhile, further experiments indicate that the proposed method shows higher trajectory accuracy and effectiveness in comparison with common traditional methods and other deep learning networks.

The rest of this work is organized as follows. Section 4.2 reviews the related literature about motion perception, recognition, and estimation. Section 4.3 introduces the proposed methodology for human motion recognition and estimation. Section 4.4 conducts related experiments and an assembly task. Conclusions are shown in Sect. 4.5.

4.2 Related Work

Instead of working as separate systems, human operators and robots are integrated for promoting HRC with shared cooperative activities. Thereinto, how effectively the system works is highly related to how well the robots can recognize the operator's motion and perceive the human intention. In Sect. 4.2, motion perception and recognition, and motion estimation are reviewed.

4.2.1 Motion Perception and Recognition

Motion perception and recognition ensure that systems acquire humans' motion states. Existing approaches can be roughly divided into two categories in the industrial scenario, i.e., visual sensors-based motion capture device and wearable sensor-based motion capture device, as shown in Table 4.1.

In the visual sensors-based motion capture system, human motion data could be obtained by extracting key contextual information from pixels of videos or pictures [7]. However, the performance of traditional motion perception and recognition methods depends on hand-crafted feature extraction methods whereas it requires prior knowledge with subjective intention. Deep learning-based approaches overcome these limitations by automating the feature extraction process. Convolution Neural Network (CNN) has been popular in classifying raw frames owing to its simplicity and efficient computation [8]. To overcome limitation of single CNN, a novel two-stream CNN has been proposed based on the calculation of optical flow difference for every two frames in the image sequence. It demonstrates excellent recognition and prediction accuracy in assembly tasks [9]. Nevertheless, the absence

Table 4.1 Summarized literature on motion perception and recognition

Approach		Methods	Refs
Visual sensors-based methods	Traditional approach	Speeded-up robust features	[7]
	Deep learning model	Convolution neural network (CNN)	[8]
	Deep learning model	Two-stream CNN	[9]
	Deep learning model	3D CNN	[10]
Inertial sensor-based methods	Kinematic model-based approach	Data processing and warning threshold algorithms	[11]
	Kinematic model-based approach	Algorithm for computing fingers' joint angles	[12]
	Artificial intelligence-based approach	Gaussian mixture model	[13]
	Artificial intelligence-based approach	Long short-term memory (LSTM) model	[14]

of spatial–temporal information in the RGB video frames limits the performance of two-stream CNN. Thus, aiming for effectively mining more motion information, 3D CNN has been utilized to capture spatial–temporal features of action images to evaluate dangerous interactions between humans and robots along with lower recognition accuracy in comparison with two-stream CNN [10]. Visual sensors-based methods are appropriate for collaboration applications with a fixed workspace and suitable light. However, a more realistic industrial task often requires the operator to walk around different assembly areas in a narrow workspace, for example, assembly tasks in aircraft cabins [15], which constrains the application of the existing methods. As a result, it is of rising demands for a more flexible motion perception and recognition method to suits the needs from a diversified assembly environment.

Wearable sensors-based methods could monitor human motions in diverse environments, which involve multiple devices, such as armbands and inertial motion capture devices. Various human motion data could be collected and analyzed to assist humans, such as estimating hand motion patterns using muscle information from a force myography band to instruct the robot to avoid collision [16], and alerting the operator by haptic feedback obtained from a vibrotactile ring [17]. As a reference in acceleration and angular velocity, inertial sensory techniques are widely adopted by various fields that involve HRC, such as construction industry, industrial robot manufacturing, and aerospace manufacturing [18, 19]. Kinematic model and deep learning model are often used to represent the inertia-related data and analyze human motions. A two IMUs motion capture system has transferred quaternions into real-time joint angles of trunk enabling workers to perceive and manage of risk factors [11]. Not only body segments' motion, but also hand movements are considered thanks to the inertial sensors. IMU-based glove ensures that the built kinematics model has calculated the relative angle of finger joints, which provides a safe working environment for collaboration [12]. Gaussian Mixture Model has used motion data to anticipate user motion intention and motion direction for the convenience of controlling robot [13]. A variety of operation motions can be recognized using the deep learning model when monitoring workers in HRC [14]. On the contrary, deep learning approaches strongly relies on massive training data and high processing power to optimize the model.

4.2.2 Motion Estimation

Motion estimation allows designing interaction plan and robotic control strategy from a holistic view in the context of changeable operation preferences and flexible tasks. Based on different motion perception devices, various techniques are applied to estimate human motion, as shown in Table 4.2.

In the visual sensors-based motion estimation, not only for single-frame input, but also for multi-frame input, bottom-up approaches and top-down approaches are frequently used. The former detects joints without a person detector, for example, posture detection using OpenPose [20] to offer support for mapping from human

Table 4.2 Summarized literature on motion estimation

Approach		Methods	Primary components	Refs
Visual sensors-based methods	Bottom-up approach	OpenPose	Multi-stage CNN	[20, 21]
	Bottom-up approach	PifPaf	ResNet, PIF, PAF	[22]
	Top-down approach	CPN	GlobalNet, RefineNet	[23]
	Top-down approach	Novel approach	Clip tracking network, video tracking network, spatial–temporal merging procedure	[24]
Inertial sensor-based methods	Traditional approach	Triaxial attitude determination method	Triaxial attitude determination method	[25]
	Traditional approach	Threshold-based method	Quaternion measurement step, Kalman Filter (KF), vector selector	[26]
	Traditional approach	Model-based method	Gauss-Markov model, linear KF	[27]
	Traditional approach	Extended Kalman Filter (EKF)	Extended Kalman Filter position and orientation constraints	[28]
	Deep-learning based approach	IONet	Bidirectional long short-term memory (Bi-LSTM)	[29]
	Deep-learning based approach	Elman neural network (ENN)-based model	ENN	[30]
	Deep-learning based approach	Hybrid network	ResNet, Bi-LSTM	[31]
	Deep-learning based approach	Recurrent neural network (RNN)-based model	Gated recurrent unit, RNN	[32]

body's ergonomic measures to robotic response [21]. PifPaf has formed full human pose using association field, which solves partial occlusion problem existing in OpenPose [22]. However, limited computational resources may not support multi-person poses' estimation in bottom-up approaches. In contrast, top-down approaches achieve better performance owing to setting a local region for each person and allowing the single-person pose detector to focus on the local region [33, 34]. In the HRC safety control system, Cascade Pyramid Network (CPN) has been designed to detect human key points and constructed human motion model [23]. To address issues of occlusion, clip tracking and video tracking networks are proposed and merged by a spatial temporal merging procedure to predict the poses of person instances not localized [24]. However, it is unable to model the spatial–temporal and visual dynamics of human joints. For the top-down approach, it is troublesome to determine proper selection modules from detection results due to the lack of compactness in bounding boxes [35]. In short, it is feasible to apply rough pre-attention and focus on the interactor for visual sensors-based motion estimation.

Wearable sensors-based motion estimation methods fall into two groups, traditional approaches and deep learning-based approaches. In traditional approaches, orientation estimation approaches are widely adopted with static and non-magnetically disturbed conditions. However, the estimation accuracy is influenced by external accelerations and ferromagnetic disturbances [25, 36]. To tackle the above issues, filter-based methods are proposed including threshold-based methods and model-based methods. The performance of the former depends on rules and heuristics, e.g., the experience of a domain expert [26]. The latter relies on constructing an accurate multi-order model [27]. In addition, the auxiliary information is proven to be beneficial to further improve the filter performance, such as position and orientation constraints [28], a combination of the skeleton model with kinematic constraints or knowledge of motion [37]. Specifically, these methods cannot handle constrained motion robustly. Deep-learning approaches have already permeated into this domain. IONet is the first deep neural network for estimating motion trajectories using raw IMU data directly [29]. It modelled the tracking problem as a sequence-based physical model and used Bidirectional Long Short-Term Memory (Bi-LSTM) for estimating walking trajectory. Recent research has filled this gap, such as human arm motion tracking using Elman Neural Network (ENN) [30], wrist motion tracking based on a hybrid network integrating ResNet and Long Short-Term Memory (LSTM) [31]. However, mentioned research only takes a few joint motions into consideration whereas multiple body segments' motion estimation is preferred to support the decision-making in HRC. For example, real-time estimation of 3D positions of body joints [32]. Deep-learning-based approaches work out solutions to problems occurring in traditional approaches, such as finding parameters and appropriate models manually, improving the versatility of models.

How to tackle the complexity of the working environment, high dimensionality in joint motions and practicability of solution are challenges in motion perception and estimation, which remain to be addressed. Visual sensors-based methods largely rely on consistent image tracking, which may result in failures under extreme lighting or camera blockage conditions. Moreover, a high-frequency image processing process leads to privacy implications and makes power consumption a bottleneck for sustainable long-term operations [38]. In contrast, wearable sensors-based methods show superiority in terms of convenience and applicability. Wearable sensor data processing requires less computational resources. Traditional methods for wearable sensors-based methods contribute to estimating human motion, accompanying by unavoidable and complicated modeling processes. Above all, the process of adjusting filter parameters and specific application scenarios hinders traditional methods popularized in many common applications. Deep learning approaches overperform the filter-based approach by automating the feature extraction with consideration of long-term temporal dependencies [39]. In this study, we go beyond the existing deep learning approaches and put more efforts on boosting wearable sensors-based device's performance by using an attention-based deep learning approach for inertial data.

4.3 Methodology

Assembly tasks often involve challenging operations in the dynamic and continuous environments, such as handling multiple devices simultaneously, moving heavy parts, and performing mobile operations. Robots can proactively assist operators in human-centered assembly tasks, i.e., HRCA. When taking an assembly task as a human–robot collaborative task, it is necessary to plan and allocate sub-tasks to the human and the robot to complement the strength of humans and robots. Significantly, human copes with situations using cognitive and flexible decision-making while robot lacks these characteristics. Intelligent algorithms and an integrated platform provide the robot with the ability to improve operational consistency with human. The proposed framework of HRCA includes the inertial motion capture device and a collaborative robot for the assembly task, as shown in Fig. 4.1, which consists of three layers, i.e., physical layer, interaction layer, and application layer. Physical layer contains robot and sensing devices, which provides the capabilities of collecting and transmitting human motion data between the inertial motion capture device and the robot. As the base layer of the framework, it is equipped with multiple software systems, such as ROS, Tensorflow. Information layer involves the analysis of sensing data and control of robot. It exploits the hardware or platforms in the physical layer to implement kinematics algorithms and load deep learning models for realizing motion perception, recognition, and estimation. Moreover, with the triggered commands, information layer is also responsible for robotic motion plan. Application layer shows the integrated system for HRCA on the basis of the physical layer and the information layer. According to the recognized and estimated human motion trajectories, the robot executes a motion plan duly for assisting operators.

4.3.1 Task Description and Modeling

Preprogrammed control commands require robots and humans to execute assembly tasks in order. Operation preferences and changeable assembly procedures may lead to failures in HRC. In this work, perception, recognition and estimation of human motion are accomplished, which help the robot to follow collaboration strategy and adjust motion states in response to human behavior. Taking a power module assembly in the laboratory as an example, several screws and connection wires need to be

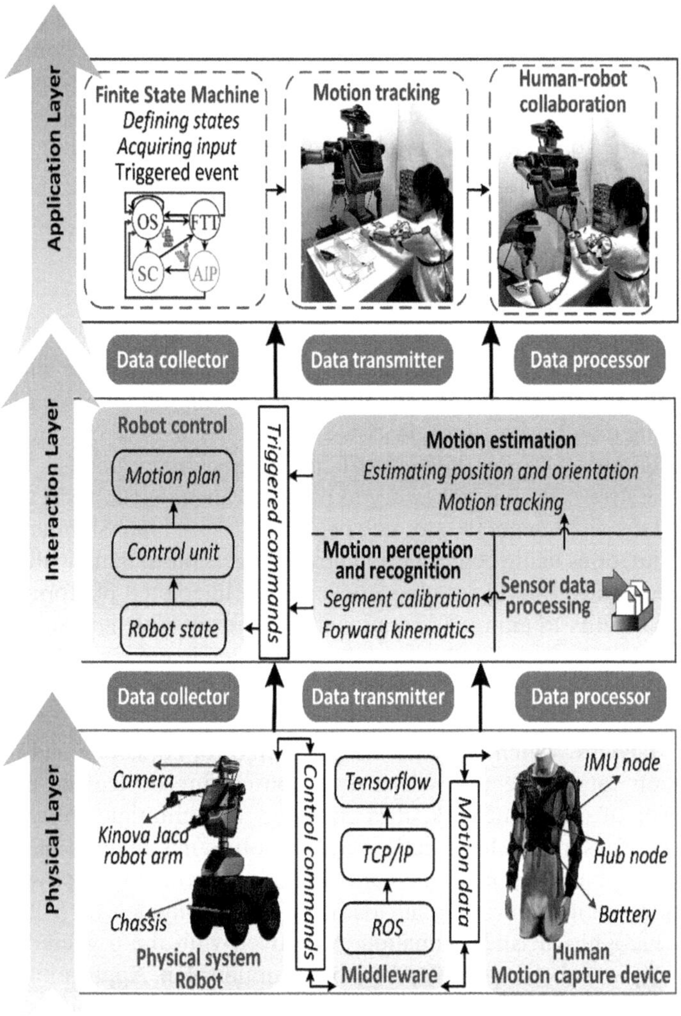

Fig. 4.1 Structure of the proposed framework of HRCA

installed. Fixed installation processes of screws and connection wires are divided into several steps.

(1) Installing screw: picking up a screw and inserting it in a hole first, spinning and adjusting, and then, picking up a screwdriver from toolbox, screwing it in, and putting the screwdriver back.
(2) Installing connection wire: picking up a connection wire, twining the connection wire around a screw first, and then, picking up a screwdriver from toolbox, screwing it in, followed by putting the screwdriver back.

HRCA is related to different states of human and robot. Results of motion perception, recognition and estimation become trigger of driving robot, which models a Finite State Machine (FSM) [40] with the following robot's states:

(1) On Standby (OS): robot is on standby at the initial position when human does not step towards targeted part. OS state may be influenced by different human states and environment settings.
(2) Fetch Targeted Tool (FTT): robot arrives at the position for fetching a targeted tool, which results from the existence of human motion's success in getting the targeted parts. When stepping toward the targeted tool, robot's motion is not interrupted except for emergencies or reset command.
(3) Arrive Interaction Position (AIP): robot arrives at the position for interacting with human when human uses the targeted part to execute assembly task. AIP state always goes after FTT state unless human stops assembling.
(4) Start Collaboration (SC): robotic grippers are controlled to assist human in assembly tasks, and human operator focuses on interacting with robot in the meanwhile. Delivery of assembly tools is achieved during SC state.

The proposed FSM is shown in Fig. 4.2. First of all, human operator is ready for executing task and robot keeps OS state. When human gets a targeted part used for assembly task, robot transmits to FTT state and fetches a targeted tool. If human executes assembly operation with the targeted part, robot changes from FTT state to AIP state. Conversely, if human does not execute assembly task though possessing a targeted part, robot will end with OS state. After robot arrives at AIP state, its next state relies on human's interaction intention. If human motion is classified as intending to interact with robot, the robot will transmit to SC state. Otherwise, robot will transmit to FTT state, put the tool back, and end with OS state. When completing SC state, robot transforms current state to OS state, then, human continues assembly task using the targeted tool and puts the tool back until human state changes to OS state. Moreover, OS state is the final state of other states when emergencies or reset command are triggered.

4.3.2 *Proposed Methodology*

Perception means the acquisition and pre-processing of raw motion data, and recognition indicates reconstructing human motion using processed data to approximate human motion states. Subsequently, with the estimated human body segments' accelerations, angular velocities, and orientations, tracking motion trajectory is realized using a deep learning model in the part of motion estimation. The acquired feature map facilitates to get the adaptive feature refinement and learn the relationship between motion data and position more effectively using the proposed attention mechanism. Estimated results facilitate to execute FSM in HRCA for the convenience of controlling robot to assist human.

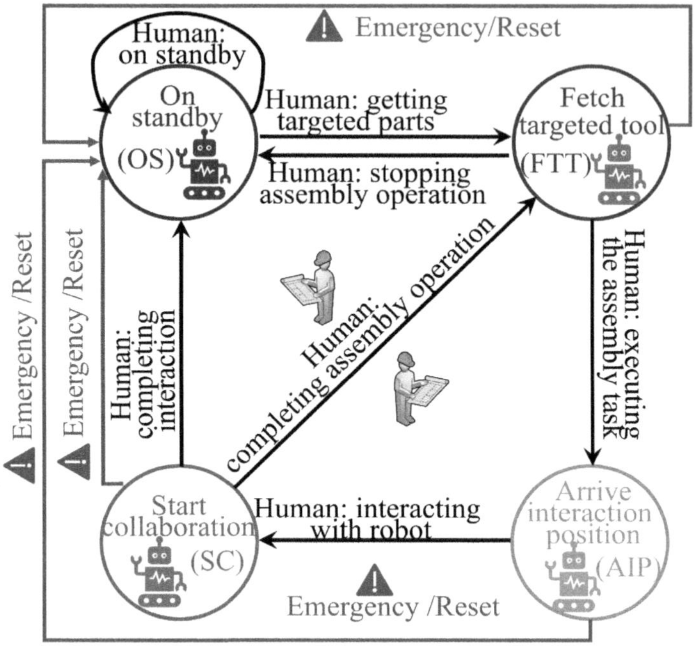

Fig. 4.2 FSM transition map for HRCA

4.3.2.1 Motion Perception and Recognition

Motion perception acquires raw motion data from the developed inertial motion capture system. Motion recognition devotes to acquiring the estimated orientation of whole body segments based on sensor-to-body segment calibration, and further uses the prior kinematics knowledge of human body to reconstruct human motion.

The developed inertial motion capture device consists of IMU nodes and a hub node for acquiring human motion data. Many IMU nodes are attached to human body segments for reading acceleration, angular velocity, magnetic field, and the hub node is used to obtain motion data from all IMU nodes and transmit data to the host computer wirelessly. Motion data from IMU nodes only represents the motion states of IMU in the sensor coordinate frame while reconstruction of human motion requires body segments' orientations. Sensor coordinates are not aligned with body segment coordinates. Therefore, it is important to calibrate the transformation orientation between the sensor coordinate frame and the body segment coordinate frame, i.e., sensor-to-body segment calibration, with the purpose of computing orientations of body segments.

Quaternion-based senor-to-body segment calibration algorithm is applied in the research. Quaternion is adopted to analyze rotations of IMU nodes and body segments. The original output from IMU nodes, i.e., q_s (quaternions of IMU nodes), represents the relationship between the sensor coordinate and the global coordinate.

The targeted orientation for reconstructing human motion is q_b that indicates the body segment coordinate relative to the global coordinate. The essence of senor-to-body segment calibration algorithm is to obtain the relationship between the sensor coordinate and the body segmt coordinate, i.e., q_{trans} that is not changed during motion perception. It is assumed that the orientation of the body segment coordinate coincides with the global coordinate when a human performs the upright calibration pose at time t_0,

$$q_{s_i}^{t_0} \otimes q_{trans_i} = q_{b_i}^{t_0} = \begin{bmatrix} 1 \, 0 \, 0 \, 0 \end{bmatrix} \tag{4.1}$$

$$q_{trans_i} = q_{s_i}^{t_0 *} \otimes q_{b_i}^{t_0} \tag{4.2}$$

Based on the calibration procedure, body segment's orientation at time t_n, $q_{b_i}^{t_n}$, can be acquired,

$$q_{b_i}^{t_n} = q_{s_i}^{t_n} \otimes q_{trans_i} = q_{s_i}^{t_n} \otimes \left(q_{s_i}^{t_0^*} \otimes q_{b_i}^{t_0} \right) \tag{4.3}$$

where $q_{b_i}^{t_0}$ and $q_{s_i}^{t_0}$ represent orientations of i_{th} body segment relative to body segment coordinate system and i_{th} sensor relative to global coordinate system at t_0 respectively, $q_{b_i}^{t_0}$ and $q_{s_i}^{t_0}$ are obtained by performing standing upright calibration pose, $q_{b_i}^{t_n}$ and $q_{s_i}^{t_n}$ represent orientations of i_{th} body segment relative to body segment coordinate system and i_{th} sensor relative to global coordinate system at t_n respectively, q_{trans_i} denotes the transformation between i_{th} sensor coordinate system and i_{th} body segment coordinate system, * refers to the conjugation of quaternion, $\otimes$ refers to the product of quaternions.

The whole human body skeleton's motion is driven by combining the motion of each body segment. In detail, the human pose can be recon structured by connecting body segments' vectors v_{b_i} based on forward kinematics when the pelvis is chosen as a root node. According to the estimated orientation, the real-time body segment's vector $p_{b_i}^{t_n}$ in the body segment coordinate system is determined by combining the orientation and vector of body segments. Each body segment's position relative to the global coordinate system $P_{b_i}^{t_n}$ is computed as below:

$$p_{b_i}^{t_n} = q_{b_i}^{t_n} \otimes v_{b_i} \tag{4.4}$$

$$P_{b_i}^{t_n} = \sum_{m=0}^{i} p_{b_m}^{t_n} \tag{4.5}$$

where $p_{b_i}^{t_n}$ is the position of i_{th} body segment relative to its parent body segment at t_n, $q_{b_i}^{t_n}$ represents orientation of i_{th} body segment relative to global coordinate system at t_n, v_{b_i} is the vector of i_{th} body segment relative to its parent body segment, $P_{b_i}^{t_n}$ is the position of i_{th} body segment in the body segment coordinate system while the position of the body segment coordinate system's origin is assumed to be static at t_n.

In HRCA, when the operator executes assembly task after completing the upright calibration pose, reconstructed human motion acquired by motion perception and recognition can reflect the motion state in the body segment coordinate system, more precisely, estimated orientations and estimated positions of body segments are obtained, which is used to promote motion estimation.

4.3.2.2 Motion Estimation

(1) Overview of Network. With the processed motion data after motion perception and recognition, the deep learning method combining residual modules, convolution modules and attention module is leveraged to estimate motion trajectory. In the proposed network, residual modules [41] detect initial features based on sequential inertial data (human body segments' accelerations and angular velocities). What follows is to get expanded features except for initial features by using the attention module to expand the dimensionality of input elements. Subsequently, the attention module measures and weighs processed input elements (i.e., initial features and expanded features) quantitatively based on the Efficient Channel Attention (ECA) block [42]. Bi-LSTM is then adopted to estimate the relative pose while the trajectory estimation is acquired based on successive changes of orientations. Finally, fully connected layers generate the relative orientation and relative displacement for acquiring absolute orientation and absolute displacement.

(2) Deep learning network structure. With the acquisition of human body segments' motion data by motion perception and recognition, they are then transmitted to the deep learning model whose details are shown in Fig. 4.3. Firstly, the processed inertial data of the reading window is transmitted into the convolutional module including batch normalization and Rectified Linear Units (ReLU) activation. Inertial estimations of different axes are concatenated and convolved using one-dimensional convolution operation with kernel size 15 and filters 96, and then construct a feature map, which is followed by a residual learning framework consisting of four residual modules. The residual module is common in the similar research [31], which is implemented with double-layer skips containing one-dimensional convolutional layers, nonlinearized operation, and batch normalization in between. Meanwhile, skip connection keeps the feature space of input using max pooling and convolution. Convolutional filters and output size of four residual modules are set differently, i.e., the parameters of the first residual module are 128 and 256, parameters of the other three are 196 and 128, 256 and 64, 320 and 32, correspondingly. The kernel size of all convolution operations is 3 and the activation function is ReLU.

Furthermore, in order to improve the performance of the network, where and what to focus on are also important after feature extraction of residual modules.

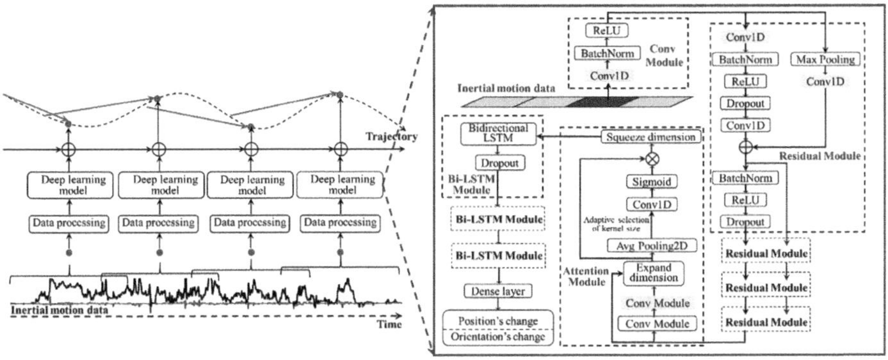

Fig. 4.3 Structure of the proposed deep learning-based network

Thus, attention module lying between residual modules and Bi-LSTM copes with the aforementioned problems, which expands the dimensionality of input data for learning more contextual knowledge, and then, inputs the raw data and the expanded data into ECA module. Concretely, attention model first adopts convolution modules to acquire the convolutional features of input, and then per forms the dimension expansion based on the initial input features and the processed features. The function $\mathcal{F}_1$ shows the convolution module to be executed, which denotes convolution, batch normalization and ReLU activation. It constructs a similar identity mapping of one-channel input rather than a direct mapping of one-channel input, which facilitates the model to find the more features. The operation $\oplus$ is performed by a shortcut connection and dimension expansion, which uses the output of $\mathcal{F}_1$ and the original input x to construct y_1. The function $\mathcal{F}_2$ employs the ECA module that increases little model complexity and suffers from lighter computational burden while comparing with SE Net [43], CBAM [44], AANet [45]. Local cross-channel interaction is paid attention to by taking each channel and its neighbors into account. The attention module brings clear performance gain by learning expanded channel attention. At last, dimensionality reduction strategy, i.e., function $\mathcal{F}_3$, extracts the aggregated initial feature efficiently while abandoning the processed aggregated convolutional feature. The following operation "squeeze" aims to decrease dimensions. In conclusion, given the one-dimension input feature of inertial signal, the proposed attention model guides the process of reasoning and captures multi-channel.

$$y_1 = \mathcal{F}_1(x, \{W_1\}) \oplus x \tag{4.4}$$

$$y_2 = \mathcal{F}_2(\sigma(C_1(pool(y_1)))), k = \left| \frac{\log_2 C}{\gamma} + \frac{b}{\gamma} \right| \tag{4.5}$$

$$y_3 = \mathcal{F}_3(y_2, \{W_3\}) \tag{4.6}$$

where x is the input of attention module, $\mathcal{F}_1$, $\mathcal{F}_2$, and $\mathcal{F}_3$ indicate operations in attention module, W_1 represents related parameters of $\mathcal{F}_1$, σ is a Sigmoid function, C_1D is 1D convolution, kernel size k is determined by channel dimension C, γ and b are determined based on experience, W_2 represents related parameters of $\mathcal{F}_3$.

Trajectory estimation involves long-term dependencies, therefore, when representing a single timestamp, the correlation among the long sequence of input data should be considered. Bi-LSTM consists of forward LSTM and backward LSTM for exploiting the previous forward information and the future backward information as well [31]. The stacked Bi-LSTM model solves complex localization problems by considering the past and future inertial features. Thus, it takes processed features as input and processes them recursively to produce the hidden state followed by a dropout layer for avoiding overfitting. The number of units of forward LSTM and backward LSTM is 96. A dropout layer with the rate of 0.25 is applied to eliminate parameters and help the model adapt to unseen data. The final step is to generate relative displacement and relative orientation using fully connected layers.

The output of the deep learning model represents relative orientation and displacement between continuous timesteps. The absolute displacement p_t and absolute quaternion q_t at time t are calculated based on relative orientation and displacement from the deep-learning network of motion tracking, i.e., Δp and Δq, and displacement p_{t-1} and quaternion q_{t-1} at previous timestep $t-1$.

$$p_t = p_{t-1} + R_{t-1}\Delta p \tag{4.9}$$

$$q_t = q_{t-1} \otimes \Delta q \tag{4.10}$$

where R_{t-1} is the corresponding rotation matrix of q_{t-1}, $\otimes$ represents the product of quaternions.

(3) Loss function.

Loss function is designed by the differences of position and orientation between the ground truth and estimated results. Three-dimension positions and quaternions are chosen as representations, which are not suitable to use algebraic metrics for evaluating the proposed network's performance. The metrics, $Loss_{pos}$ and $Loss_{qua}$, are shown below,

$$Loss_{pos} = \left\|\widehat{pos} - pos\right\| \tag{4.11}$$

$$Loss_{qua} = 2\left\|imag(\widehat{qua} \otimes qua^{*}\right\| \tag{4.12}$$

where pos represents the estimated position, $\widehat{pos}$ is the ground truth of the position, qua represents the estimated quaternion, and $\widehat{qua}$ is the ground truth of quaternion, $\|\cdot\|$ represents the L2 normalization, and $imag$ shows the imaginary part of variable.

The final goal is to find the optimal parameters by minimizing the loss term. Instead of assuming average weight of losses, the composite loss function, $Loss$, tries to find the optimal weight for every loss based on multi-task learning method, as follows,

$$Loss = \sum_{i=1}^{2} \exp(-\log \sigma_i^2) Loss_i + \log \sigma_i^2 \tag{4.13}$$

where σ_i shows the weight of loss function of every task, $Loss_i$ represents metrics of task, i.e., $Loss_{pos}$ and $Loss_{qua}$.

4.4 Case Study and Experiments

In this section, a case study and related experiments are conducted to confirm the advantages of the proposed approach. First, the operator performs an assembly task of a power module with different parts, and robot helps human by delivering assembly tools for tightening operation. Then, to verify the effectiveness of our proposed model, algorithms' performances are shown and compared with other algorithms based on the benchmark dataset and our own dataset. Finally, experimental results are discussed.

4.4.1 An Illustrative Human-Robot Collaboration Assembly Task

From the perspective of HCPS, HRC consists of three major parts, i.e., human, cyber system, and physical system. Human operator wearing the developed motion capture device undertakes manual assembly tasks. Manual operation motion is acquired through IMUs and a cross-system interface for transmitting data between human and robot. Human completes calibration pose and random poses in turn according to instructions. All human motion data is collected and processed in real time, which is embedded in the application. Robot equipped with two Kinova Jaco robotic arms is controlled based on the ROS control framework. It takes on assistance during interaction with human in an assembly task. Data interface transmits human motion data from inertial motion capture device to robot. Thereinto, Transmission Control Protocol/Internet Protocol (TCP/IP) and ROS computation graph mechanism ensure real-time motion data transmission from inertial motion capture device to robot. Data interface extracts raw data and processed data for collecting and estimating human behaviors. Similarly, deep learning model developed by Tensorflow is integrated with ROS for motion estimation. The assembly task is represented as serial-parallel operations, where human and robot perform the operation procedures in sequence. It involves some typical operations during the task. Firstly, collaborative robot, assembly parts, and assembly tools are placed at the expected location, and

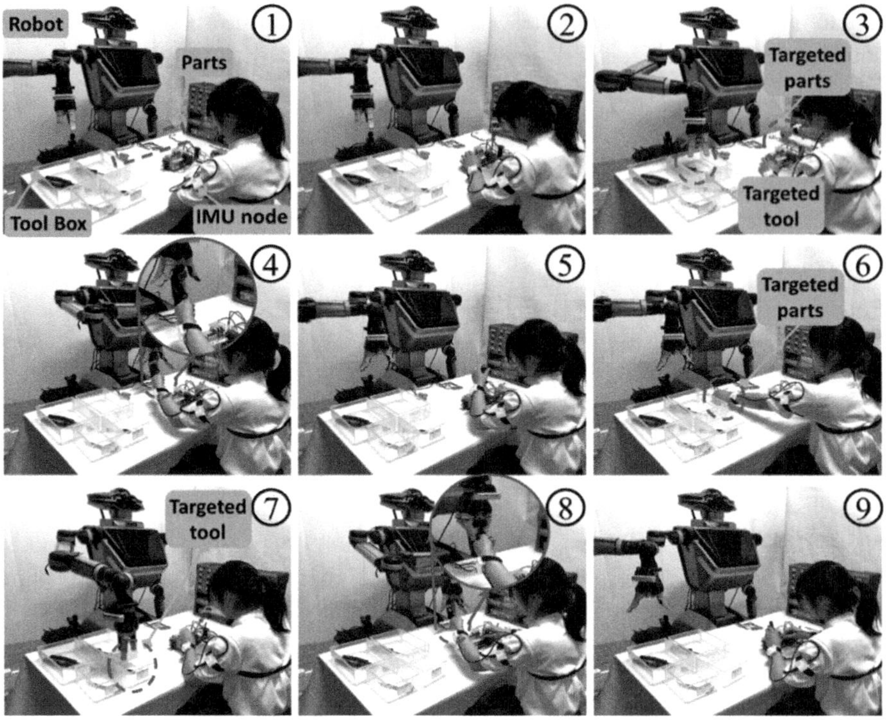

Fig. 4.4 Process of an assembly task case

the operator gets ready for assembly task, as shown in Fig. 4.4(1). Then, operator starts to execute assembly task in Fig. 4.4(2), i.e., she picks up a screwdriver and unscrews a screw. When the screw is loose, operator whose motion is acquired and estimated takes a targeted part for installation (Fig. 4.4(3), (6)). Robot picks up the targeted tool when the estimated result of human motion meets the requirements of robot control based on FSM framework (Fig. 4.4(3), (7)). The delivery of the targeted tool between robot and operator promotes the assembly task (Fig. 4.4(4), (8)), and operator ends the task by tightening screws for fastening targeted parts (Fig. 4.4(5), (9)).

4.4.2 Experiments of Motion Perception and Recognition

An experiment based on the benchmark motion capture system has been executed, which compares the estimated human pose using the proposed perception and recognition method with the human pose from the benchmark system.

The complex setup and susceptibility to lighting conditions and occlusion have an effect on the optical motion capture system or visual sensors-based motion capture system. Therefore, the benchmark inertial motion capture system (Perception Neuron, Noitom, China) was adopted as the ground truth of trajectory tracking [46]. Participants were recruited to wear the developed inertial motion capture device and the benchmark inertial motion capture device simultaneously, and then executed different poses. Figure 4.5 shows human motion in the physical system and virtual human model motion in the cyber system synchronously. In order to validate motion perception and recognition intuitively, the matched graphical user interface is built for the developed inertial motion capture device (i.e., the left virtual human model in Fig. 4.5(1)–(6)). At the same time, the right virtual human model is ac quired from the benchmark system directly. Moreover, in order to assess the correlation between two methods, further motion analysis is executed on comparing measured joint angles from two systems. We extracted motion data of dynamic movement, i.e., leg changes from the outstretched state (Fig. 4.5(5)) to the bending state (Fig. 4.5(6)). Joint angles are calculated using estimated human body segments' orientations from the proposed algorithm and the benchmark algorithm, which are applied to analyze correlation. Table 4.3 reports the results of correlation analysis and agreement analysis, and the former result is evaluated using Pearson Correlation Coefficient (PCC) while the latter uses the confidence probability based on Bland–Altman analysis. PCCs of the out stretched state and bending state are equal or larger than 99%, which shows a strong positive correlation between the proposed algorithm and benchmark algorithm. Similarly, confidence probabilities acquired from Bland–Altman analysis are larger than 98%, which reveals the agreement between the two algorithms. In short, the proposed method shows good performance in terms of motion perception and recognition in comparison with the benchmark method.

In an assembly task, operator takes on the assembly task primarily and handles assembly parts and assemble device manually. It is beneficial for knowing the approximate range of operator current working state by motion perception and recognition. Once real-time arm motion state meets requirements of state's transition in FSM, HRC system starts to record human motion data for estimating arm motion trajectory.

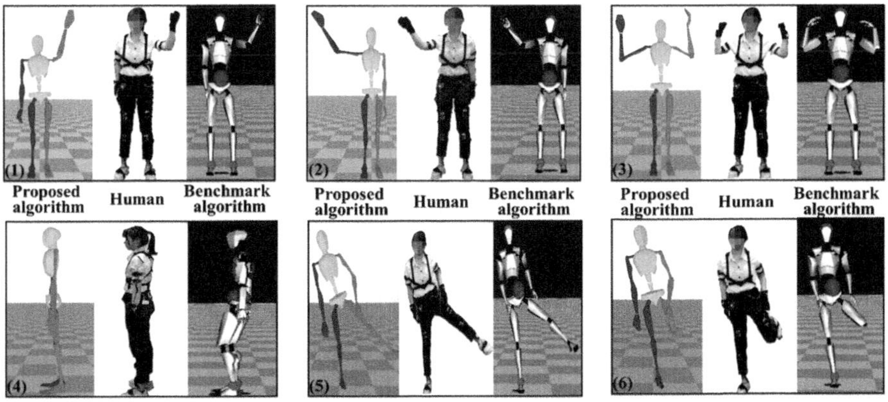

Fig. 4.5 Display of physical human motion and virtual model motion

Table 4.3 Results of correlation analysis and agreement analysis between the proposed algorithm and the benchmark algorithm

Motion state	Pearson correlation coefficient (%)	Confidence probability (%)
Outstretched state	99.1	98.6
Bending state	99.0	98.5

4.4.3 Experiments of Motion Estimation

The proposed deep-learning model estimates human upper body motion in a predominantly human–robot collaborative environment, as the prerequisite to ensure HRC. A benchmark motion dataset is first leveraged to evaluate the performance of the proposed method when compared with traditional methods and other deep learning-based methods. Then, the other dataset created by the developed motion capture device is utilized to demonstrate the performance for facilitating the applicability of the proposed network. The major metrics include mean error, median error, maximum error, Absolute Trajectory Error (ATE) and Relative Trajectory Error (RTE). The first three metrics (i.e., mean error, median error, maximum error) represent the error between the estimated motion trajectory and the corresponding ground truth trajectory directly. ATE calculates the RMSE between the predicted motion trajectory and the ground truth from the perspective of the whole process. RTE depicts RMSE between the estimated trajectory and the ground truth trajectory for a fixed time interval.

4.4.3.1 Model Performance with Different Architectures

Before making a decision on the optimal structure, different architectures, and parameters are thought about to select the optimal model. Experiments present the performance of combinations of different structural parameters using self-collected data such as the number of residual modules, bidirectional/unidirectional LSTM, and the number of LSTMs. Table 4.4 shows the performance of combinations of different Residual Modules for feature detection and extraction, there into, numbers of Residual Modules and sizes of input/output vectors are paid much attention to. Five configurations of the model, i.e., 1–5 Residual Modules are tested when rest parts remain the same. It can be found that four Residual modules with 256 * 128 (input vector's size * output vector's size), 128 * 196, 64 * 256, 32 * 320 on each layer respectively achieve the best performance. Similarly, the attention module plays a vital role in the model, therefore, the proposed attention mechanism is compared with a similar structure without adding the attention module, i.e., Res-ECA network. It leverages a stack of residual modules, a Bi-LSTM module, and a ECA module. The inertial data is processed by residual modules firstly, and input into the original ECA module without expanding features. The Bi-LSTM module uses the output data from the ECA module to estimate motion trajectory. As shown in Table 4.5, the proposed network's performance is superior to the Res-ECA network. At last, Table 4.6 shows the performance of different layers for bidirectional LSTM and unidirectional LSTM. Results demonstrate that the three-layers bidirectional LSTM outperforms other structural combinations. In conclusion, the optimal structure combining four residual modules, the proposed attention module and three layers of Bi-LSTM is used for the further comparison of different parameters setting.

Except for structural design, experiments related to parameters are executed, including activation function, learning rate, lookback size, and batch size. Common settings of related parameters are considered, i.e., four common activation functions, four learning rates, and four batch sizes. The size of lookback relies on sampling frequency while the sampling frequency of two devices is 200 Hz by sampling preprocessing, thus, four lookback size are compared. Table 4.7 presents comparing the performance of models with different parameters by using the optimal structure

Table 4.4 Comparison of residual modules with different structure

Residual module/layers	Size of input vector and output vector of each layer	X axis/mm		Y axis/mm		Z axis/mm		Sum/mm	
		ATE	RTE	ATE	RTE	ATE	RTE	ATE	RTE
5	512 * 64, 256 * 128, 128 * 196, 64 * 256, 32 * 320	40.7	28.7	45.1	29.4	39.6	31.4	125.4	89.5
4	256*128, 128 * 196, 64 * 256, 32 * 320	36.4	23.7	40.9	23.6	46.4	31.4	123.7	78.7
3	128 * 196, 64 * 256, 32 * 320	35.5	21.3	52.0	32.2	39.1	27.6	126.6	81.1
2	64 * 256, 32 * 320	39.4	24.4	53.7	32.3	43.0	30.0	136.1	86.7
1	32 * 320	39.8	23.9	55.0	31.8	41.5	29.3	136.3	85.0

Table 4.5 Comparison of different networks in terms of attention module

Model structure	Attention module/layers	X axis /mm		Y axis /mm		Z axis /mm		Sum/mm	
		ATE	RTE	ATE	RTE	ATE	RTE	ATE	RTE
Res-ECA network	0	35.1	20.7	48.1	31.5	47.6	34.5	130.8	86.7
Proposed network	1	38.3	24.8	42.9	24.5	37.4	26.2	118.6	75.5

Table 4.6 Comparison of LSTM and Bi-LSTM with different structure

Module	Layers	X axis/mm		Y axis/mm		Z axis/mm		Sum/mm	
		ATE	RTE	ATE	RTE	ATE	RTE	ATE	RTE
LSTM	1	36.4	25.1	45.3	29.1	41.0	28.6	122.7	82.8
	2	35.4	22.8	50.2	31.2	38.9	26.8	124.5	80.8
	3	35.8	22.1	50.4	30.7	46.0	32.4	132.2	85.2
	4	38.7	24.2	51.1	31.8	40.7	28.6	130.5	84.6
Bi-LSTM	1	34.8	22.8	45.9	28.0	38.9	26.9	119.6	77.7
	2	38.0	24.3	40.6	25.2	39.2	26.5	117.8	76.0
	3	34.8	21.8	42.2	24.7	38.1	26.5	115.1	73.0
	4	40.1	25.8	40.6	22.4	38.8	26.4	119.5	74.6

selected from previous experiments. Experiments indicate that the optimal results are obtained when the optimization algorithm is Nadam, learning rate is 0.001, lookback size is 500, and batch size is 128.

Table 4.7 Comparison of performance with different parameters using the proposed network

Parameters		X axis/mm		Y axis/mm		Z axis/mm		Sum/mm	
		ATE	RTE	ATE	RTE	ATE	RTE	ATE	RTE
Activation function	SGD	317.0	221.5	380.9	262.7	276.3	190.3	974.2	674.5
	Adam	40.7	25.1	46.0	29.2	47.1	32.7	133.8	87.0
	Nadam	37.8	24.8	43.8	26.3	37.6	26.6	119.2	77.7
	RMSProp	46.0	30.5	51.6	31.5	40.4	27.6	138.0	89.6
Learning rate	0.01	51.3	34.1	56.1	34.8	50.8	34.9	158.2	103.8
	0.005	49.2	33.0	45.0	26.4	46.4	32.0	140.6	91.4
	0.001	37.8	24.8	43.8	26.3	37.6	26.6	119.2	77.7
	0.0005	37.1	25.1	45.1	25.5	44.2	30.6	126.4	81.2
Lookback size	700	48.5	25.2	67.3	32.5	55.3	27.2	171.1	84.9
	500	37.8	24.8	43.8	26.3	37.6	26.6	119.2	77.7
	300	46.4	24.8	49.5	31.7	48.9	31.6	144.8	88.1
Batch size	256	42.2	28.3	47.3	28.7	42.6	28.9	132.1	85.9
	128	37.8	24.8	43.8	26.3	37.6	26.6	119.2	77.7
	64	39.9	24.8	50.2	31.7	45.2	31.6	135.3	88.1
	32	35.9	22.1	51.8	30.6	43.9	31.2	131.6	83.9

4.4.3.2 Evaluation on the Benchmark Dataset

The benchmark dataset recorded a healthy 28 years old male's upper body motion which involves one function degree of freedom at a time. The motion data were captured using inertial motion capture system, camera-based motion capture system and Vicon optical motion capture system. All motion sequences last at least 10 s, which are processed and gathered at 100 Hz. Each sequence of movements was performed at low speed and high speed for testing methods' performance [47]. Yang has used this benchmark dataset for validating the performance of ENN-based model when compared with traditional methods. In order to evaluate the performance of the proposed method, motion estimation experiments are carried out based on the experimental results of ENN-based model [30]. Four traditional methods (i.e., Zhu's model [48], Yun's model [49], Young's model [50], and Bleser's model [51]) and five deep learning-based methods (i.e., ENN-based model, hybrid network, CNN-LSTM model, Res-ECA model, the proposed model) are evaluated using the benchmark dataset.

Table 4.8 and Fig. 4.6 show experimental trajectory errors of positions from traditional methods and deep learning-based methods. The results of traditional methods come close to each other in terms of errors of X-axis, but errors of Y-axis and Z-axis differ considerably. For example, Z-axis's mean error of Yun's model is almost six times higher than Young's model. It is shown that the performances of different traditional methods vary dramatically. In contrast, different deep learning-based methods report similar performance, moreover, they all demonstrate obvious improvements for motion tracking in comparison with traditional methods as a whole, except for metrical results of Z-axis from some deep learning-based methods, for example, three metrics of Z-axis belonging to ENN-based model show worse trajectory accuracy when comparing with Zhu's model. Experimental results of the proposed model and CNN-LSTM model show the proposed model's superiority in X-axis and Y-axis except for Z-axis. In order to further evaluate deep learning-based methods, ATE is also used to measure accuracy. Figure 4.6b shows ATE results among four deep learning models. It shows consistency with the metrical results of errors in Table 4.8, i.e., the proposed method gets higher motion trajectory accuracy from the perspective of X-axis and Y-axis except for Z-axis. Motion trajectory represents motion in three-dimension space, therefore, it is necessary to make a comparison using accumulative errors of three axes rather than only one axis. In brief, the proposed method shows better performance in comparison with other methods when taking accumulative metrical results of three axes into consideration. Crucially, it is obvious that the variant of the proposed model, i.e., Res-ECA model, shows a worse performance than the proposed model. Therefore, the better performance of the proposed model depends not only on the hybrid architecture including residual model and Bi-LSTM but also on the proposed attention module.

Table 4.8 Errors from traditional methods and deep learning-based methods on the benchmark dataset

Method		X-axis/mm			Y-axis/mm			Z-axis/mm		
		Mean error	Median error	Maximum error	Mean error	Median error	Maximum error	Mean error	Median error	Maximum error
Traditional methods	Zhu's model	38.7	31.7	128.1	88.3	90.5	237.5	17.8	13.7	85.3
	Yun's model	54.7	49.7	148.4	50.0	45.3	188.7	75.3	60.8	267.3
	Young's model	40.8	31.1	136.2	83.2	89.0	206.7	13.5	9.0	69.4
	Bleser's model	45.6	39.0	156.1	39.2	34.4	187.7	39.1	31.6	157.8
Deep learning-based methods	ENN-based model	24.3	18.9	114.4	19.7	17.4	65.5	50.3	30.7	220.4
	Hybrid network	27.1	16.1	98.9	21.3	10.3	73.7	25.7	8.5	114.3
	CNN-LSTM model	29.8	23.3	101.7	21.4	14.8	70.2	18.4	9.6	68.1
	Res-ECA model	23.3	15.7	101.2	19.3	10.5	68.2	21.4	12.5	75.0
	Proposed model	21.4	11.3	94.3	15.8	9.3	69.4	19.6	9.3	75.5

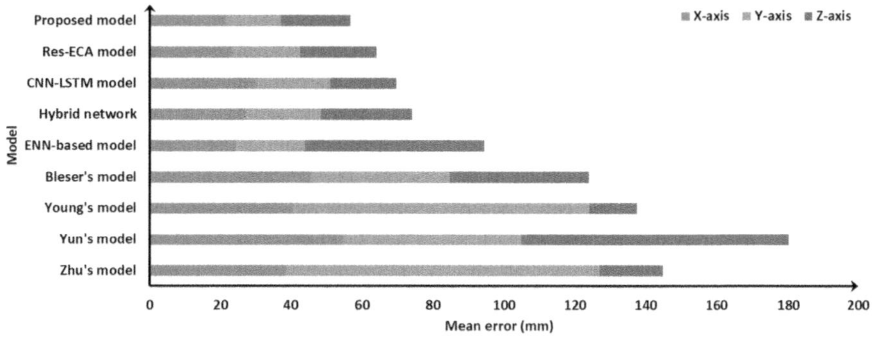

(a) Evaluation results of mean errors from traditional methods and deep learning-based methods

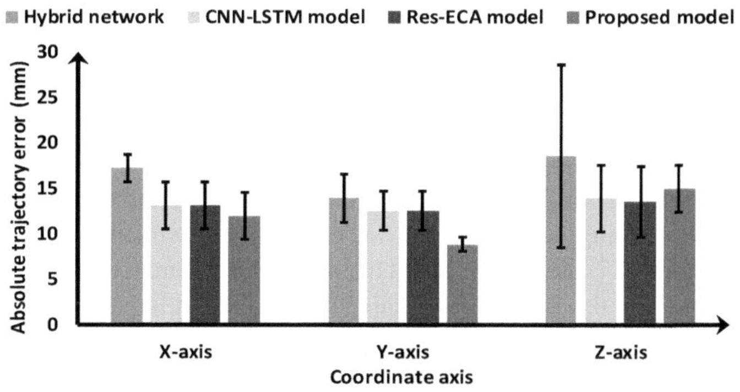

(b) Evaluation results of ATE from deep learning-based methods

Fig. 4.6 Evaluation results from various methods on the benchmark dataset. **a** Evaluation results of mean errors from traditional methods and deep learning-based methods, **b** evaluation results of ATE from deep learning-based methods

4.4.3.3 Evaluation of Our Own Dataset

Inertial motion estimation has been widely researched in various fields, and existing datasets primarily are related to pedestrian inertial navigation or vehicle navigation [38, 52]. In vehicle navigation, the dataset is built using vehicle movement data from sensors fixed to the chassis of a car, which is not suitable for analyzing human movements. Similarly, dataset for pedestrian inertial navigation records different motion modes, such as halting, walking, and running. It is also not appropriate for analyzing arm motion. Upper body motion is frequently involved during human–robot interaction. Our own dataset recording human arm motion trajectory is developed because the benchmark dataset possesses a small amount of data and a few types of motion trajectories.

Our dataset collects data of IMUs for training and evaluating the deep learning-based models. The benchmark inertial motion capture system, Perception Neuron, was adopted as the ground truth of the trajectory tracking [46]. In order to record upper body motion, all IMU nodes for capturing whole body motion are used whereas only upper motion data is applied to train and test. The raw motion data of the developed inertial motion capture system and benchmark inertial motion capture system, and the processed positions and orientations of benchmark inertial motion capture system are read and collected through ROS. It is considered that different operations vary in motion trajectories resulting from assembly environment and operation experience, for example, assembly parts in different positions lead to different arm motion trajectories when human operator tries to take it. Thus, two types of motion trajectories are performed including short-distance movements and long-distance movements. Short-distance movements include "—" shape, "|" shape, "L" shape while long-distance movements include "Δ" shape, "Z" shape. Corresponding lengths of each type of motion trajectory are 0.5 m, 0.5 m, 0.7 m, 2.0 m, and 2.4 m approximately. Experiments with different motion trajectories were performed by participants with two inertial motion capture devices after participants execute calibration poses. In the training dataset, each type of motion trajectory has 30 sequences, and each sequence represents a kind of whole motion trajectory.

The raw motion data and the processed data are fed into four deep learning models, i.e., hybrid network, CNN-LSTM model, Res-ECA model, and the proposed model. Note that these training motion trajectories are not present in the testing motion trajectories. Figure 4.7 shows five motion trajectories estimated from the proposed method (red line) and the ground truth trajectories (blue line) meanwhile. A series of experiments are conducted to compare different deep learning methods in different motion trajectories. Tables 4.9 and 4.10 and Fig. 4.8 show evaluation results measured by ATE and RTE. The proposed method performs better localization accuracy than other approaches in these metrics on our own dataset in X-axis and Y-axis while its performance does not outperform other models in Z-axis. CNN-LSTM model works well on Z-axis's motion estimation among all deep learning methods. Errors of motion trajectories from the hybrid network become slightly large, which presents that the hybrid network may not account for the orientation difference and displacement difference accurately. Most notably, Res-ECA model cannot handle motion tracking well in comparison with the proposed model, which concludes that the proposed attention mechanism module improves ATE and RTE overall. It shows that the proposed method produces estimation results closer to the ground truth in short-distance movements and long-distance movements. Figure 4.7 shows the estimated motion trajectories from the proposed method and the ground truth. This experimental phenomenon on our own dataset stays in step with results on the benchmark dataset.

Motion trajectory represents positions in three-dimensional space, therefore, metrics from the sum of three axes are considered, just shown in Tables 4.9 and 4.10. The proposed model performs better localization accuracy than other approaches in these metrics on our own data set. ATE and RTE acquired using the proposed model are 117.4 mm and 77.2 mm respectively. ATEs of the proposed model are 12.6, 14.9, 7.3% lower than that of the hybrid network, CNN-LSTM model and Res-ECA model. RTEs of the proposed model are 12.6, 12.4, 3.9% lower than that of the hybrid

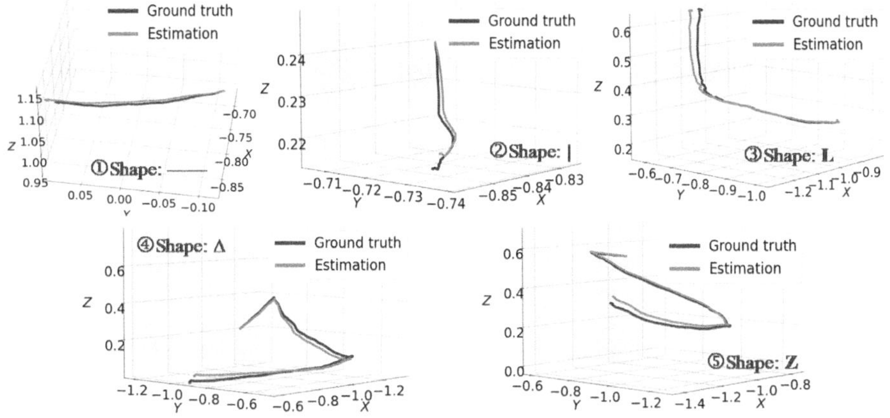

Fig. 4.7 Estimated motion trajectories from the proposed method and the ground truth motion trajectories

Table 4.9 Errors from deep learning-based methods on our own dataset

Model	X-axis/mm		Y-axis/mm		Z-axis/mm		Sum/mm	
	ATE	RTE	ATE	RTE	ATE	RTE	ATE	RTE
Hybrid network	41.8	26.4	47.5	30.1	45.1	31.8	134.4	88.3
CNN-LSTM model	46.5	30.8	50.2	28.8	41.3	28.5	138.0	88.1
Res-ECA model	34.8	20.3	44.6	26.9	47.3	33.1	126.7	80.3
Proposed model	34.3	21.6	37.4	23.4	45.7	32.2	117.4	77.2

Table 4.10 Improvements over our own dataset of the proposed model in comparison with other deep learning-based methods

Model	X-axis/mm		Y-axis/mm		Z-axis/mm		Sum/mm	
	ATE (%)	RTE (%)	ATE (%)	RTE (%)	ATE (%)	RTE (%)	ATE (%)	RTE (%)
Hybrid network	17.9	18.2	21.3	22.3	− 1.3	− 1.3	12.6	12.6
CNN-LSTM model	26.2	29.9	25.5	18.8	− 10.7	− 13.0	14.9	12.4
Res-ECA model	1.4	− 6.4	16.1	13.0	3.4	2.7	7.3	3.9

network, CNN-LSTM model and Res-ECA model. Not only ATE but also RTE of motion trajectories are improved using the proposed model. In short, the proposed model outperforms competing approaches to different extents.

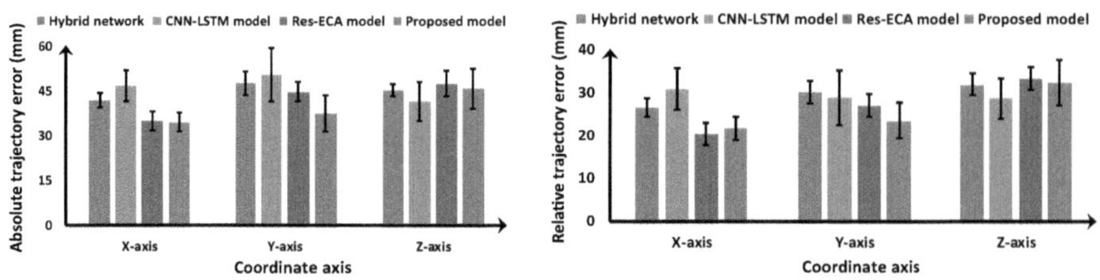

(a) Evaluation results of ATE (b) Evaluation results of RTE

Fig. 4.8 Evaluation results from deep learning-based methods on our own dataset. **a** Evaluation results of ATE, **b** evaluation results of RTE

4.5 Conclusion

To facilitate wearable device-based motion recognition and estimation in HRCA, an attention-based deep learning approach has been developed. It applies inertial motion data to acquire current human motion state and estimate human motion trajectory, so that the robot determines when and how to aid human worker. Two aspects are taken into consideration, i.e., motion perception and recognition, motion estimation. In the motion perception and recognition, body segments' orientations are acquired and reconstructed through sensor-to-body segment calibration and kinematics analysis. It has achieved comparable performance with the benchmark system, as reflected in the 99% agreement analysis and 98% correlation analysis. In the motion estimation, using benchmark dataset and our own dataset, the incorporation of convolution module, residual module, attention module and Bi-LSTM has improved the predicted human motion trajectory accuracy as compared to traditional methods and existing deep learning methods. In particular, the proposed attention module based on squeeze-and-excite mechanism and channel attention is effective in improving ATE and RTE of motion trajectories in comparison with Res-ECA model, by up to 7.3% ATE and 3.9% RTE. A case study that cites an assembly task as the example is implemented to verify the feasibility of the proposed approach. The proposed approach provides an insight into the industrial implementation of HRC based on deep learning and wearable device. Understanding human's working state and manufacturing environment, as well as their interaction, appears rewarding for HRCA.

With respect to future work, it will extend collaborative applications of the proposed method, which are highlighted here: (1) integrating vision sensor with inertial sensor for facilitating human motion perception and estimation; (2) empowering intuitive and efficient human–machine interface using augmented reality/virtual reality.

References

1. Zhou J, Zhou Y, Wang B, Zang J (2019) Human-cyber-physical systems (HCPSs) in the context of new-generation intelligent manufacturing. Engineering 5(4):624–636
2. Wang B, Zheng P, Yin Y et al (2022) Toward human-centric smart manufacturing: a human-cyber-physical systems (HCPS) perspective. J Manuf Syst 63:471–490
3. Zhang R, Lv Q, Li J et al (2022) A reinforcement learning method for human-robot collaboration in assembly tasks. Robot Comput-Integr Manuf 73:102227
4. Wang B, Zhou H, Yang G et al (2022) Human digital twin (HDT) driven human-cyber-physical systems: key technologies and applications. Chin J Mech Eng 35:11
5. Roveda L, Maskani J, Franceschi P et al (2020) Model-based reinforcement learning variable impedance control for human-robot collaboration. J Intell Rob Syst 100(2):417–433
6. Zhou H, Yang G, Lv H et al (2020) Iot-enabled dual-arm motion capture and mapping for telerobotics in home care. IEEE J Biomed Health Inform 24(6):1541–1549
7. Rublee E, Rabaud V, Konolige K, Bradski G (2011) ORB: an efficient alternative to SIFT or SURF. In: International conference on computer vision, Barcelona, Spain, pp 2564–2571
8. Chen C, Zhao X, Wang J et al (2022) Dynamic graph convolutional network for assembly behavior recognition based on attention mechanism and multi-scale feature fusion. Sci Rep 12:7394
9. Zhang J, Wang O, Gao RX (2021) Hybrid machine learning for human action recognition and prediction in assembly. Robot Comput-Integr Manuf 72:102184
10. Amin FM, Rezayati M, Venn HW, Karimpour H (2020) A mixed-perception approach for safe human-robot collaboration in industrial automation. Sensors 20(21):6347
11. Yan X, Li H, Li AR, Zhang H (2017) Wearable IMU-based real-time motion warning system for construction workers' musculoskeletal disorders prevention. Autom Constr 74:2–11
12. Bright T, Adali S, Bright G (2022) Low-cost sensory glove for human-robot collaboration in advanced manufacturing systems. Robotics 11(3):56
13. Tortora S, Michieletto S, Stival F, Menegatti E (2019) Fast human motion prediction for human-robot collaboration with wearable interfaces. In: IEEE International conference on cybernetics and intelligent systems (CIS) and IEEE conference on robotics, automation and mechatronics (RAM), Bangkok, Thailand, pp 457–462
14. Kim K, Cho YK (2020) Effective inertial sensor quantity and locations on a body for deep learning-based worker's motion recognition. Autom Constr 113:103126
15. Simões AC, Pinto A, Santos J, Pinheiro S, Romero D (2022) Designing human-robot collaboration (HRC) workspaces in industrial settings: a systematic literature review. J Manuf Syst 62:28–43
16. Anvaripour M, Khoshnam M, Menon C, Saif M (2020) FMG- and RNN-based estimation of motor intention of upper-limb motion in human-robot collaboration. Front Robot AI 7:573096
17. Casalino A, Messeri C, Pozzi M et al (2018) Operator awareness in human-robot collaboration through wearable vibrotactile feedback. IEEE Robot Automation Letters 3(4):4289–4296
18. Ciccarelli M, Papetti A, Germani M, Leone A, Rescio G (2022) Human work sustainability tool. J Manuf Syst 62:76–86
19. Baghdadi A, Megahed FM, Esfahani ET, Cavuoto LA (2018) A machine learning approach to detect changes in gait parameters following a fatiguing occupational task. Ergonomics 61(8):1116–1129
20. Cao Z, Hidalgo G, Simon T et al (2019) OpenPose: realtime multi-person 2D pose estimation using part affinity fields. IEEE Trans Pattern Anal Mach Intell 43(1):172–186
21. Broek MK, Moeslund TB (2020) Ergonomic adaptation of robotic movements in human-robot collaboration. In: Companion of the 2020 ACM/IEEE international conference on human-robot interaction, pp 499–501
22. Kreiss S, Bertoni L, Alahi A (2019) PifPaf: composite fields for human pose estimation. In: Proceedings of the IEEE/CVF conference on computer vision and pattern recognition, Long Beach, CA, USA, pp 11977–11986

23. Li H, Ma W, Wang H et al (2022) A framework and method for human-robot cooperative safe control based on digital twin. Adv Eng Inform 53:101701
24. Wang M, Tighe J, Modolo D (2020) Combining detection and tracking for human pose estimation in videos. In: IEEE/CVF conference on computer vision and pattern recognition, Seattle, WA, USA, pp 11085–11093
25. Wertz JR (1978) Three-axis attitude determination methods. In: Wertz JR (ed) Spacecraft attitude determination and control. Springer, Dordrecht, Netherlands, pp 420–428
26. Lee JK, Park EJ (2009) Minimum-order Kalman filter with vector selector for accurate estimation of human body orientation. IEEE Trans Rob 25(5):1196–1201
27. Ligorio G, Sabatini AM (2015) A novel Kalman filter for human motion tracking with an inertial-based dynamic inclinometer. IEEE Trans Biomed Eng 62(8):2033–2043
28. Joukov V, Lin JF-S, Kulić D (2019) Closed-chain pose estimation from wearable sensors. In: IEEE-RAS 19th international conference on humanoid robots, Toronto, ON, Canada, pp 594–600
29. Chen C, Lu X, Markham A, Trigoni N (2018) IONet: learning to cure the curse of drift in inertial odometry. In: Proceedings of the AAAI conference on artificial intelligence, pp 6468–6476
30. Yang H, Wang Y, Wang H et al (2022) Multi-inertial sensor-based arm 3d motion tracking using Elman neural network. J Sens 3926417
31. Lin B-S, Lee I-J, Wang S-P, Chen J-L, Lin B-S (2022) Residual neural network and long short-term memory-based algorithm for estimating the motion trajectory of inertial measurement units. IEEE Sens J 22(7):6910–6919
32. Wei W, Kurita K, Kuang J, Gao A (2021) Real-time limb motion tracking with a single imu sensor for physical therapy exercises. In: 43rd annual international conference of the IEEE engineering in medicine and biology society, Mexico, pp 7152–7157
33. Wang X, Feng R, Chen H et al (2022) Personalized motion kernel learning for human pose estimation. Int J Intell Syst 37(9):5859–5879
34. Pavllo D, Feichtenhofer C, Grangier D,Auli M (2019) 3D human pose estimation in video with temporal convolutions and semi-supervised training. In: IEEE/CVF conference on computer vision and pattern recognition, pp 7745–7754
35. Song Z, Yin Z, Yuan Z et al (2021) Attention-oriented action recognition for real-time human-robot interaction. In: 2020 25th international conference on pattern recognition, Milan, Italy, pp 7087–7094
36. Shuster MD, Oh SD (2012) Three-axis attitude determination from vector observations. J Guid Control Dyn 4(1):70–77
37. Joukov V, Bonnet V, Karg M, Venture G, Kulić D (2018) Rhythmic extended Kalman filter for gait rehabilitation motion estimation and segmentation. IEEE Trans Neural Syst Rehabil Eng 26(2):407–418
38. Liu W, Caruso D, Llg E et al (2020) TLIO: tight learned inertial odometry. IEEE Robot Autom Lett 5(4):5653–5660
39. Chen L, Xu G, Zhang S, Yan W, Wu Q (2020) Health indicator construction of machinery based on end-to-end trainable convolution recurrent neural networks. J Manuf Syst 54:1–11
40. Ren Y, Li G-P (2022) A contextual sensor system for non-intrusive machine status and energy monitoring. J Manuf Syst 62:87–101
41. He K, Zhang X, Ren S, Sun J (2016) Deep residual learning for image recognition. In: IEEE conference on computer vision and pattern recognition, Las Vegas, NV, USA, pp 770–778
42. Wang Q, Wu B, Zhu P et al (2020) ECA-Net: efficient channel attention for deep convolutional neural networks. In: IEEE/CVF conference on computer vision and pattern recognition, Seattle, WA, USA, pp 11531–11539
43. Hu J, Shen L, Albanie S, Sun G, Wu E (2020) Squeeze-and-excitation networks. IEEE Trans Pattern Anal Mach Intell 42(8):2011–2023
44. Woo S, Park J, Lee J, Kweon I-S (2018) CBAM: convolutional block attention module. In: European conference on computer vision. Springer, Cham, pp 3–19
45. Xu H, Zhang J (2020) AANet: adaptive aggregation network for efficient stereo matching. In: IEEE conference on computer vision and pattern recognition, Seattle, WA, USA, pp 1956–1965

46. Choo CZY, Chow JY, Komar J (2022) Validation of the perception neuron system for full-body motion capture. PLoS One17(1):e0262730
47. Filippeschi A, Schmitz N, Miezal M et al (2017) Survey of motion tracking methods based on inertial sensors: a focus on upper limb human motion. Sensors 17(6):1257
48. Zhu R, Zhou Z (2004) A real-time articulated human motion tracking using tri-axis inertial/ magnetic sensors package. IEEE Trans Neural Syst Rehabil Eng 12(2):295–302
49. Yun X, Aparicio C, Bachmann ER, McGhee RB (2005) Implementation and experimental results of a quaternion-based Kalman filter for human body motion tracking. In: Proceedings of the 2005 IEEE international conference on robotics and automation, Barcelona, Spain, pp 317–322
50. Young AD (2010) Use of body model constraints to improve accuracy of inertial motion capture. In: International conference on body sensor networks, Singapore, pp 180–186
51. Bleser G, Hendeby G, Miezal M (2011) Using egocentric vision to achieve robust inertial body tracking under magnetic disturbances. In: 10th IEEE international symposium on mixed and augmented reality, Basel, Switzerland, pp 103–109
52. Chen H, Zhu H, Li M, You S (2018) A review of visual-inertial simultaneous localization and mapping from filtering-based and optimization-based perspectives. Robotics 7(3):45

Chapter 5
Conclusion and Outlook

Abstract This chapter summarizes core research advancements in human motion capture and its transformative applications for smart manufacturing. It systematically examines motion capture methodologies encompassing multimodal data acquisition, adaptive pose estimation, and predictive trajectory modeling. These techniques are integrated with teleoperated robots to enable human–robot collaboration. The concept of Phygital Twin is introduced as a paradigm-shifting framework, emphasizing bidirectional knowledge transfer between human operators and their cyber-physical avatars. Future directions prioritize three critical frontiers: (1) Based on human motion digital twin, developing resilient motion capture systems with enhanced environmental adaptability for dynamic industrial settings; (2) Leveraging AI-driven analytics for context-aware robotic responsiveness to human behavioral patterns; (3) Establishing cross-industry scalability through interoperable digital twin architectures, such as human digital twin and physical system digital twin. The discussion advocates for human-centric interface, aiming to accelerate the evolution of cognitive manufacturing ecosystems where human expertise and machine precision achieve symbiotic optimization.

Keywords Human motion digital twin · Phygital twin · Motion capture

5.1 Conclusion

In the field of smart manufacturing, human motion digital twin is emerging as a crucial technology for achieving high levels of automation and effective human–machine collaboration. This book focuses on the research and application of human motion digital twin in the context of human–robot interaction. It addresses gaps in existing research and provides valuable theoretical and technical support for advancing smart manufacturing. This book begins with an introduction to data processing technology which is foundational for creating accurate and reliable human motion digital twin systems. In smart factories, this technology is utilized to monitor and analyze human motion, optimize workstation design to enhance production efficiency and safety.

H. Zhou et al., *Revitalizing Human-Robot Interaction*, Advanced Topics in Science and Technology in China 3, https://doi.org/10.1007/978-981-96-6139-8_5

This book explores the development of human motion digital twin systems based on IMUs. By precisely capturing human movement, these systems offer deep insights into human behavior and operational patterns. In a smart manufacturing environment, this enables robots and automated systems to interact more naturally and intuitively with humans. In addition, the book focus on key technologies and applications related to human motion perception, cognition, and interaction. The analysis of human movements further optimizes human–robot collaboration by these technologies. The main research content of this book is as follows:

(1) In the smart manufacturing, human motion calculation and calibration are not only important for acquiring human motion states but also fundamental for developing human motion digital twins. These technologies ensure that robots and automated systems accurately interpret and respond to human motion, thereby enhancing productivity and safety. By calculating and calibrating posture, the motion state of human can be captured and analyzed into operator behavior patterns. This data supports the design of more ergonomic workstations, optimizes task distribution, and improves job satisfaction. Furthermore, accurate motion data allows human motion digital twins to simulate and predict real human movements in a virtual environment. This capability is invaluable for improving product design, conducting risk assessments, and training employees. In smart manufacturing, digital twins can be employed to test and optimize production processes, anticipate potential problems and bottlenecks, and achieve cost and time savings without physical trials.

This book proposes an extended Kalman filter algorithm for attitude calculation based on adaptive error estimation. It constructs the motion state equation and observation equation of the inertial sensor node, adaptively estimates the measurement noise covariance matrix, and iteratively solves the quaternion of the sensor node's attitude based on the vector observation prior calculation value based on the Gauss–Newton algorithm as the initial value. In terms of motion calibration, this book proposes the human motion calibration algorithm based on human kinematic constraints. It clarifies the transformation relationship between limb segment posture and sensor node posture, obtains the quaternions of the main limb segments of the human body, and reconstructs human motion. Furthermore, the above algorithm was integrated into a self-developed inertial motion capture system, and the RMSE of the calculated static posture joint angle of the human body was 4.451°, while the RMSE of the dynamic motion joint angle was 4.700°. It showed a strong positive correlation with the joint angles obtained based on optical motion capture systems and had good consistency, verifying the feasibility of the system.

(2) Human motion mapping and planning not only enhance the intuitiveness and interaction of human–machine interfaces but also significantly improve the accuracy and efficiency of remote operations. Accurately capturing the operator's motion intentions is crucial for effective remote operation. In a smart factory, this involves real-time, precise tracking of minute operator movements, allowing robots or remote devices to replicate these actions for

complex assembly, commissioning, and maintenance tasks. Designing an effective motion mapping plan is vital to enhancing the naturalness and intuitiveness of remote operation interactions. In smart manufacturing, such plans must address the complexity and diversity of operations to ensure that the system accurately understands and executes the actions of various operators. Additionally, the mapping plan should be flexible to adapt to different task requirements and individual differences.

This book constructs a human coordinate system, calculates the position vector of the end joint of the human arm, and estimates the position vector of the end joint of the arm. In response to the demand for human–machine motion mapping interaction, this book designs a pose mapping strategy based on the human–machine workspace and imitates the motion characteristics of human arm. It improves the humanoid nature of the robot arm's motion following. The experimental results show that the model proposed in this paper achieves good accuracy in tracking the trajectory of arm-end joint motion, which is superior to other comparative models. At the same time, by integrating the above algorithms into the hardware system, the laboratory remote operation interaction application and the China-Sweden intercontinental remote operation interaction application have been completed.

(3) Efficient human–machine cooperation is crucial for enhancing production efficiency and competitiveness in the smart manufacturing. This requires accurate estimation of human motion and the development of effective coordination strategies to optimally allocate human and machine resources. By constructing models of human behavior, smart factories can better integrate the flexibility of human workers with the efficiency of robots to foster a complementary work environment. Human motion estimation technology captures and analyzes worker movements, providing data-driven insights to optimize workstation layouts, reduce ineffective movements and fatigue. Additionally, this data supports the development and adaptation of automation systems and robot control systems, ensuring seamless collaboration with human colleagues. Designing effective collaborative strategies involves merging human intuition, decision-making ability, and flexibility with machine persistence, precision, and efficiency. This needs to develop algorithms and protocols that guide robots in responding to human actions, selecting optimal modes of collaboration. Moreover, establishing human behavior models allows smart manufacturing systems to predict and adapt to changes in human behavior. This capability helps identify potential conflicts and bottlenecks in advance.

This book combines human pose calculation methods and multi-channel attention deep learning models to predict human motion trajectories. The channel attention mechanism is used to focus on multi-channel features, extract original and extended features, and estimate the output pose through a sequence model. Furthermore, this book establishes a human–machine interaction state transition model based on finite state machines, simulates assembly scenarios in a laboratory environment. Human–machine collaborative assembly is treated as an example to promote application analysis, verifying the feasibility of the above methods. The experimental results show that the human motion trajectory prediction model proposed in this paper has high prediction accuracy in both short-distance and long-distance motion trajectory estimation scenarios.

5.2 Discussion[1]

5.2.1 Revitalizing Human-Robot Interaction Through Phygital Twin

Digital Twin (DT) is built and maintained in the digital rather physical realm [1, 2]. Dynamic DT mirrors the exact state of the physical system by means of multidimensional models describing the behavior of physical entity, and sensors providing the real-time coupling to models [3]. The outbreak of COVID-19 pandemic results in speeding up the virtualization and digitization of the physical entity with the support from DT. Nevertheless, quarantine and division led by COVID-19 pandemic make human beings' eyes for the physical interaction become more eager. The physical world and the digital world are becoming interwoven, which indicates a new opportunity towards phygital twin. As is so often the question: how can I get closer to my distant companion? How can I extend my service or emotional concern to facilitate interaction in spite of space obstruction? These questions deserve thinking as a lesson learned from outbreak of COVID-19 pandemic.

DT focuses on how digital elements to represent physical objects in the abstract and condensed ways based on a confluence of technological advances [4]. Intrinsic difference between human beings and physical systems leads to different emphasis in the description by DT [5]. When involved into human-centric DT, human users expect tangible and intuitive experience to meet requirements of engagement and immersion. They need to prioritize transparency and intuition providing for humans. Moreover, not only interaction enhancement, but also getting rid of the limitations of space are primary requirements of the next-generation DT in favor of human's involvement. Phygital twin emerges at this critical moment.

"Phygital" is a portmanteau of "physical" and "digital". It is the concept of utilizing technology to bridge the reality and cross-space reality by means of interaction between digital world and physical world, aiming for enhancing unique and interactive experiences of users, providing a positive outcome with a seamless toggling between the physical realm and the digital realm.

Phygital twin is a term allowing trio among physical reality, digital world, and the other cross-space physical reality. Its connection changes the paradigm by unifying

---

[1] Revised from Zhou et al. Revitalizing Human-Robot Interaction: Phygital Twin Driven Robot Avatar for China-Sweden Teleoperation, Chinese Journal of Mechanical Engineering, Copyright (2023), with permission from Springer Open.

Author: Huiying Zhou, Honghao Lv, Ruohan Wang, Haiteng Wu, and Geng Yang*

multiple hardware infrastructure and software infrastructure [6]. With advanced technology, phygital twin has an opportunity to endow avatar with more rational and emotional factors. Imaging taking care of distant family, sharing your joy or sorrows with distant friend, receiving warm hug from lovers, surpass verbal communication [7].

In terms of realization, robot is quite an appropriate vehicle of phygital twin thanks to its advanced technologies and broad application [8, 9]. Telerobotics is one of the earliest aspects of robotics. It refers to robotics with a human operator in control or human-in-the-loop. Furthermore, human is responsible for any high-level planning, or cognitive decisions while robot executes mechanical implementation. Researchers attempt to apply human intelligence, cognition, and skills to robots. Selvaggio et al. introduced the first haptic-guided shared control teleoperation architecture [10]. It is used in the transportation of an object on a tray-like end-effector. It allows integrating human operators' experience and cognitive capabilities with the precision, reactivity, and repeatability of autonomous systems. In addition to the control strategy, the transparency and stability are also challenging. Feizi et al. proposed a real-time frequency-based delay compensation approach to maximize transparency while reducing the activation of the stabilization layer [11]. However, human–robot teleoperation is the epitome of phygital twin in terms of technology scale while phygital twin does not just refer to operation of systems or machines at a distance. Zhang et al. presented a mixed-reality avatar attached with robot in the dynamic interaction among humans, robot, and mixed-reality avatar robot [12]. Beik-Mohammadi et al. integrated reinforcement learning with the model mediated teleoperation. It can handle challenging problems in a long-distance teleoperated grasping scenario under long time-delays [13]. In comparison with telerobotics, researchers related to phygital twin strive to create an environment for maximizing the value of time, transcending space, and extending humans' ability. No matter how far away you are, where you are, and who you are, robot becomes your avatar to do what you want to do without needing to be there in person. Meanwhile, real-world experiences and virtual world experiences are cherished for humans. Thus, robot avatar is expected to perceive real-world, thereby facilitate human beings to sense real world and virtual world without constraints of place.

In this work, researchers step further on transcontinental phygital twin system. Section 5.2.2 introduces the system architecture of the proposed system, and Sect. 5.2.3 shows validation case study. Conclusion is given in Sect. 5.2.4.

5.2.2 New Paradigm-Phygital Twin

Phygital twin focuses on applying knowledge and abilities gained from human beings to phygital avatar, i.e., transferring information from humans with subjective initiative to physical avatar (e.g., robot). The integrated phygital twin system involves the operator located in Sweden, and the robot avatar in China (Fig. 5.1). A set of wearable devices is used to capture human's motion and provide feedback information. The

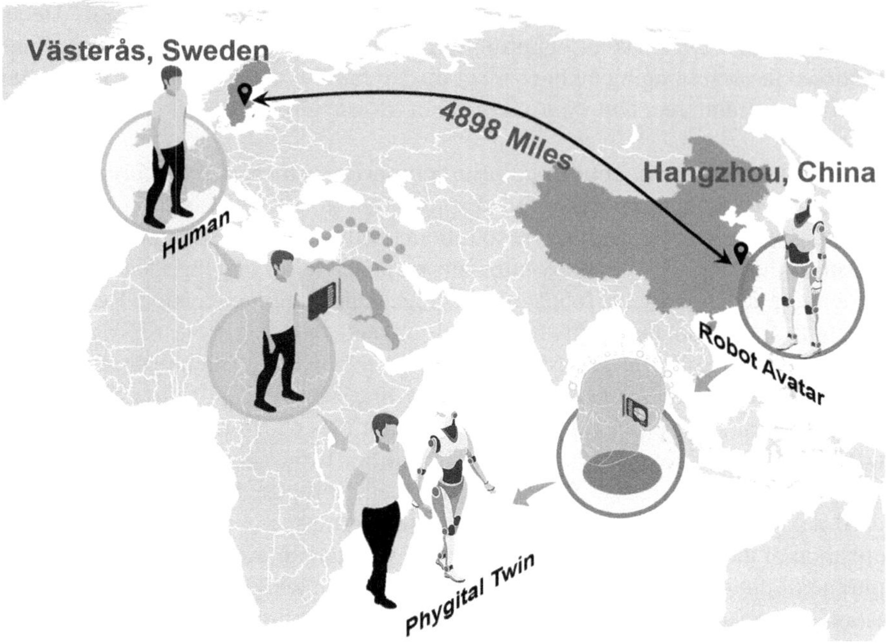

Fig. 5.1 Phygital twin system for China-Sweden teleoperation

robot receives transmitted information, replacing human for executing tasks in a first-person perspective.

(1) Interactive Interface

Interactive interface supports information transmission between the natural and the artificial side in the human–robot interaction. Intuitiveness, functionality, and affordability are taken into consideration in the context of concerning humans' well-being and rights. The well-known and traditional interface devices, such as touch joystick and keyboard, enable an easy and portable interaction way. Nevertheless, it places cognition burden on humans. Similarly, the interfaces with a high level of autonomy (e.g., brain-machine interface) makes demands on autonomy and intelligence of interface and robots. When human beings get there, it will be a magic world experienced rarely before. Now, researchers go further by developing intuitive interface for the execution of teleoperation tasks, such as wearable devices [14]. Developed inertial motion capture device (InMoCap) is employed for human motion acquisition. The wearable feedback device tracks robotic gripper openness and reflects degree of proximity sensing by robot. Wearable devices with vibrator and Light Emitting Diode (LED) allow to translate robotic grippers' operation states into feedback as a reference for teleoperating grippers.

Using IMUs, InMoCap measures primary body segments' motion. This device is made up of IMU nodes, hub microcontroller, communication module, and battery module. An IMU node integrates a gyroscope, an accelerometer, and a magnetometer with a processing unit, and acquires acceleration, angular velocity, and magnetic field for obtaining human motion. The built-in microcontroller of IMU node reads real-time motion data and communicates with the hub microcontroller. The hub microcontroller is responsible for passing control command to each IMU node and collecting motion data of attached IMUs. Embracing by multiple operation systems and application software environment, corresponding data interface is developed to receive data based on Transmission Control Protocol/Internet Protocol (TCP/IP). Figure 5.2 shows the operator and her animation acquired from InMoCap and optical motion capture system (Qualisys, Sweden). The interface is embedded into ROS for interacting with robot avatar.

(2) Dual-arm robot

Robot is an appropriate choice to become human avatar in phygital twin. They are superior in executing repetitive tasks, performing independent and fast actions in structured or known environments. Based on advanced robotics technologies, researcher proposes the future potential of "toward surrogate of human: robot avatar", allowing transferring human beings' skills to remote place. With robot avatar, interaction space where operator completes manipulation tasks acquires extension.

Robot represents the avatar of human operator in the physical system. Two six degrees of freedom KINOVA JACO2 robotic arms face to face in slightly V-shaped angle to imitate human arms' configuration. Moreover, the design of robotic arms

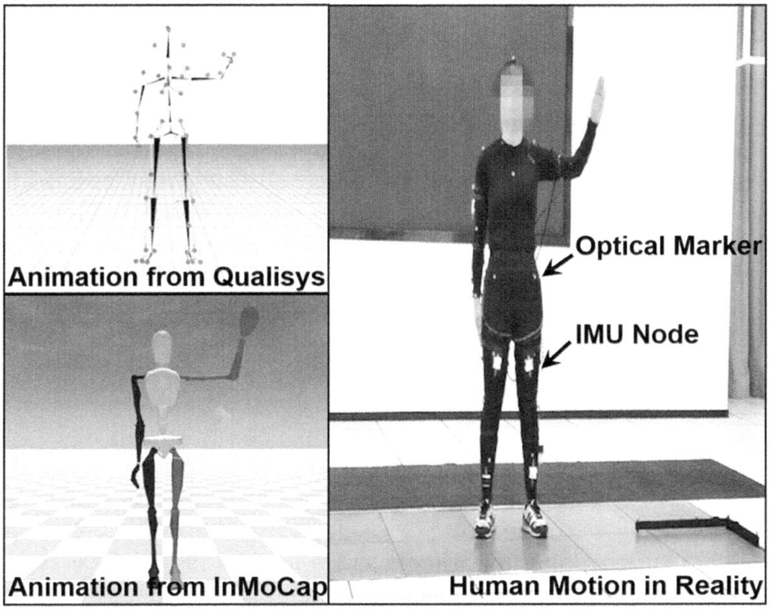

Fig. 5.2 IMU-based human motion capture and visualization

contributes to smoothness and versatility of fully functioning human arm. Each arm is capable of suffering the maximum continuous payload of 1.6 kg, the maximum workspace of 900 mm, and the maximum gripper force of 40 N with power consumption of 25 W. Each arm joint has torque sensor for force and torque measurements close to robotic arms. In order to enrich operation information, mounted depth cameras are perpendicular to robotic grippers. The head of robot is endowed with functions of images capturing and video streaming using one stabilized camera. It is placed at the middle position between two robotic arms by mimicking human head configuration. High-definition image transmission system motivates real-time video transmission with ultra-low time delay and remote transmission. It provides robot with high-performance visual ability. Robot avatar's upper body is mounted on a holonomic platform with four wheels and laser radar. It enables positioning, navigation, obstacle avoiding, and omnidirectional movement.

(3) Human-robot interaction

Phygital twin envisions a borderless, interactive, and fully immersive realm capable of covering avatars in the virtual world. Beyond virtual domain, can we extend the world based on existing border? Thus, it proposes the "reconfiguration of world" in the view of humans and their physical avatars in the physical world and virtual world. It consists of human, the physical system, and the cyber system [5].

Human operator is responsible for the perception, decision, interaction, and control. Physical robot becomes avatar of humans and promotes human–robot interaction. With sensor fusion algorithms and body segment calibration, orientations of human body segments are obtained. Thereinto, hip is regarded as the root node of the human global coordinate system. With the known body segments' orientations and constructed skeleton models, position and orientation of human hands in the human global coordinate system are mapped to the robotic end-executors by coordinate transformations between human and robot [15]. Converted motion data is transmitted to robot and leads to path planning of robotic arms. As for human–robot communication, InMoCap and robot connect to respective Local Area Networks (LANs) via wired or wireless methods. Each sub-network is controlled by sub-router respectively and made connection by building virtual private network. ROS is chosen as the unified platform for communication, control, and development. Operation station is represented as a ROS node to communicate with robot avatar by sharing topics and services. It ensures transmission of interactive data and services between human and robot.

5.2.3 Validation

The phygital twin goes after releasing potential abilities of the avatar system and combining strengths of the operator and the robot together. Inspired by robot avatar, the phygital twin system is validated in a transcontinental demonstration involving

Table 5.1 Measured metrics of IMU-based motion capture system using optical system as standard

Joint angles				Motion trajectories
Static poses	Dynamic movements			Dynamic movements
RMSE[a]	RMSE	PCCs[b]	PCT[c]	RMSE
3.9°	4.7°	92.5%	97.8%	1.2 cm

[a]*RMSE* Root mean square error; [b]*PCCs* Pearson correlation coefficients; [c]*PCT* The percentage of out of the confidence interval

the human operator and the robot system. Human wearing the InMoCap located in Västerås, Sweden, controls the robot system located in Hangzhou, China.

Effective and reliable human motion and robotic motion lay the foundation for collaborative interaction. Characteristics of InMoCap are shown in Table 5.1. The computed human motion data have been measured using the optical system as golden standard. The RMSE of joint angles in the static poses is 3.9° while is 4.7° in the dynamic movements. The Pearson correlation coefficient of dynamic motion data is about 92.5%, and the average agreement certainty of the overall joint angles is about 97.8%. RMSE between IMUbased computed motion trajectories and standard values is about 1.2 cm.

With the mapped motion trajectories from human operator, robotic arms replace human arms in the operation tasks, for example, placing blocks into container (Fig. 5.3a). The active robotic arm takes on the primary part of task, i.e., picking up the target and placing it down. By moving container, the passive robotic arm collaborates with the active arm in the placement of the target. Furthermore, dual-arm operation ability shows humans' strength in interaction. Robot system approaches targeted objects and executes task with the collaboration of two robotic arms. As shown in Fig. 5.3b, robot avatar fetches bamboo slip on the desk, and the other robotic arm fetches bamboo tube. Human operator needs to coordinate two arms to put the bamboo slip into the tube. In the operation tasks, basic feedback devices are equipped with operators. For example, when operator teleoperates the robot, robotic perception information is transmitted to operator. Specifically, operator needs to know if the robotic grippers open or not according to results of grabbing tasks. With the acquired information from vibration and flash, direct feedback information helps operator make decision conveniently.

5.2.4 Summary

This chapter presents preliminary reflections and discussions on phygital twin as a novel human–robot interaction paradigm, including preliminary framework, components, and validation case. Existing human–machine interaction paradigm is at the transitional stage where multiple enabling technologies are provided for extension

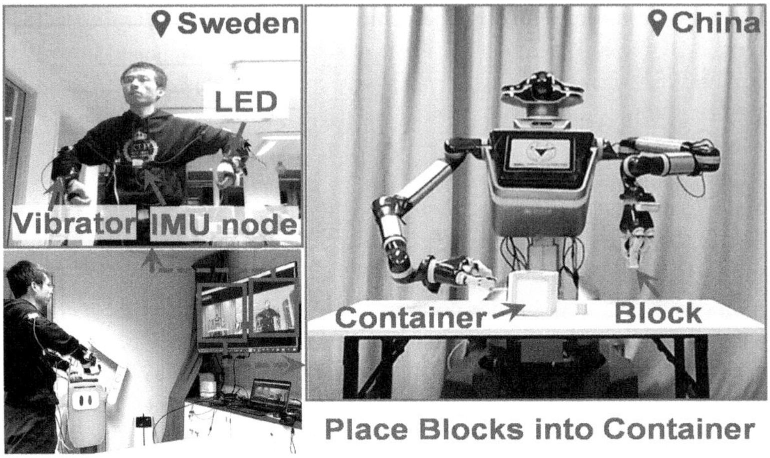

(a) Placing blocks into container

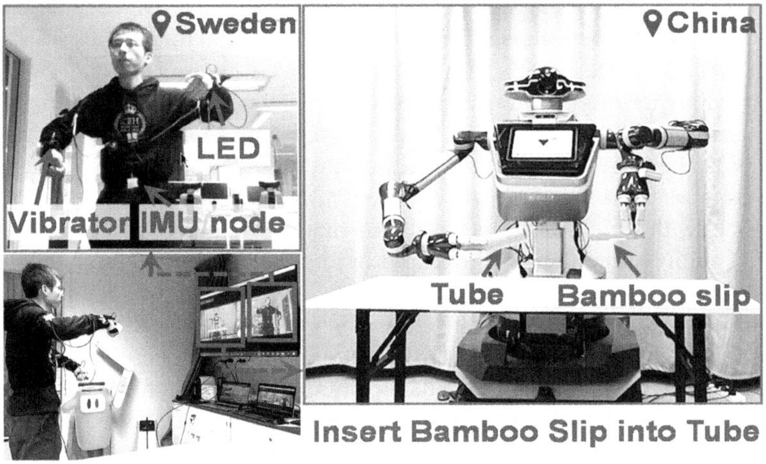

(b) Inserting bamboo slip into tube

Fig. 5.3 Transcontinental operation tasks based on Phygital Twin

and enrichment of related conceptual framework. In this paper, phygital twin facilitates interaction between human and robot avatar. Thereinto, robot avatar replaces human to execute multiple tasks with providing more sense of perception, feedback, and cognition. In particular, operation cases have been completed using transcontinental teleoperation. With increasing awareness of bridging the physical reality and cross-space physical reality, it is foreseeable that phygital twin will overthrow traditional ideas and join in next-generation human–machine paradigm.

Challenges remain in realizing and promoting phygital twin system. To promote and popularize phygital twin, it is critical to clarify a comprehensive and basic framework for implementing phygital twin in varying industrial levels. Despite all the optimism around phygital twin, it is known that we are at the early stage of adopting this vision in multiple areas. It is expected to embrace the future development across the broader application areas, such as robotics, cognitive computing, and artificial intelligence. However, there are many challenges remaining to solve, such as the availability of data from which to build phygital twin system, safety issues related to the integration of robot avatar into human-centric system. Moreover, the development of phygital twin also requires the legal and political support. The growing awareness of the vision of phygital twin will certainly bring more discussions about handling with challenges and promoting phygital twin.

5.3 Future Work

This book undertakes a comprehensive exploration of the key technologies and applications pertaining to human motion digital twins based on wearable devices. It focusses on the theoretical underpinnings and methodological frameworks, while also meticulously analyzing and discussing the integration of these systems within practical applications. Concurrently, the preceding discussions have not only introduced but also provided preliminary reflections and deliberations on the Phygital Twin, a new human–robot interaction paradigm.

Phygital Twin emerges as a powerful enabler of interaction between human and robot avatars. This innovative concept allows the robot avatar to step in and execute a diverse range of tasks on behalf of humans. It can augment the overall experience with enhanced perception, feedback, and cognitive capabilities. Notably, this has been exemplified through the successful completion of operation cases using transcontinental teleoperation, which not only showcases the technological feasibility but also highlights the potential for global collaboration and resource optimization.

As our understanding and awareness of bridging the physical reality and cross-space physical reality continue to expand, it becomes increasingly evident that the digital twin concept is a central role in the next-generation human–machine paradigm. However, the path towards realizing and promoting the Phygital Twin system is challenging.

To effectively promote and popularize the Phygital Twin, it is important to establish a comprehensive and fundamental framework that can be implemented across diverse industrial levels. This framework must take into account the unique requirements and constraints of each industry. Despite the positive optimism surrounding the Phygital Twin, it is crucial to acknowledge that we are still in the early stages of adoption in multiple areas. Future research will need to focus on addressing several key challenges. One of the primary areas will be the availability and quality of data required to build and sustain the Phygital Twin system. This will involve the development of advanced data acquisition techniques, as well as the implementation of robust data management and analytics strategies. Additionally, ensuring the safety and security of the integration of robot avatars into human-centric systems will be a top priority. This will necessitate the development of fail-safe mechanisms, rigorous testing procedures, and the establishment of industry standards and regulations. Another critical aspect that demands attention is the legal and political support required for the widespread adoption of the Phygital Twin. This includes the development of clear legal frameworks that govern liability, privacy, and ethical considerations. Politically, there will be a need for government initiatives and incentives to encourage research and development, as well as the creation of an enabling environment for innovation and commercialization. Looking ahead, future research directions in the field of Phygital Twin will span multiple disciplines. In robotics, efforts will be directed towards enhancing the dexterity, adaptability, and autonomy of robot avatars. This will involve the development of advanced control algorithms, the integration of artificial intelligence and machine learning techniques, and the exploration of novel materials and actuator technologies. In the realm of cognitive computing, research will focus on improving the understanding and interpretation of human intentions and emotions. This will enable more natural and intuitive interactions between humans and robot avatars, thereby enhancing the overall user experience. Additionally, the development of cognitive architectures that can support real-time decision-making and learning in complex and dynamic environments will be a key area of investigation. Artificial intelligence will play a pivotal role in the future of Phygital Twin. This will include the application of deep learning techniques for tasks such as object recognition, motion prediction, and behavior generation. AI-powered virtual agents and assistants will also be developed to provide intelligent support and guidance to human operators, further enhancing the efficiency and effectiveness of human–robot collaboration. Furthermore, the integration of emerging technologies such as AR, VR, and haptic feedback systems will offer new opportunities for enhancing the immersion and interactivity of the Phygital Twin experience. These technologies will enable users to interact with virtual and physical objects in more natural and engaging ways, blurring the boundaries between the real and digital worlds.

The future research prospects of the smart manufacturing-oriented human motion digital twin are as follows.

(1) The construction of a high-fidelity, real-time human motion digital twin system involves handling diverse types of human data. As digital twin systems advance, they increasingly incorporate data closely related to human ontology. This data may be distributed across various devices, such as clients, servers, production terminals, and transfer stations during application, necessitating robust security mechanisms for the digital twin system. Data security and privacy protection have thus become critical concerns. Digital twin technology relies on collecting substantial amounts of data from entities, which may be at risk of interception or tampering during transit. If attackers gain access to this data, they could compromise systems or expose sensitive information. Therefore, the security of

data collection and transmission demands urgent attention. Even when data is successfully transferred to a data center or cloud platform, the storage environment's security must be ensured, including preventing unauthorized access and protecting against malware. Additionally, data backup and recovery strategies are essential to prevent data loss or corruption. Moreover, digital twin technology involves the collection and processing of extensive personal or sensitive data, such as personally identifiable information and health status. Privacy protection is crucial to prevent the misuse or leakage of this data to third parties. If users are concerned about data security and privacy, they may be reluctant to adopt or continue using digital twin technology. Ensuring data security and privacy is not only a legal and ethical obligation but also a key factor in building user trust and facilitating the widespread adoption of the technology. Furthermore, the implementation of digital twin technology must comply with relevant laws and regulations. To address these issues, several measures are needed to enhance the security and privacy of digital twin data. These measures include using encryption to safeguard data during transmission and storage, implementing access control mechanisms to restrict data access, and employing anonymization and de-identification techniques to protect personal privacy. Additionally, it is crucial to develop and enforce laws, regulations, and ethical guidelines to ensure the legal and compliant use of digital twin technology.

(2) In modern society and industry, ensuring the relevance of workers and protecting their rights is a crucial responsibility. To achieve this goal, a human-centered approach can be adopted in industrial policy. This not only helps protect the rights of individuals and groups but also promotes the application of value-based policies to ensure that technological progress benefits every member of society. First, a people-centered strategy emphasizes the dignity and worth of workers, ensuring that they remain relevant and competitive in a rapidly changing technological environment. This includes providing the necessary training and re-education to enable workers to master and adapt to new technologies and ways of working. By doing so, we can avoid unemployment caused by technological progress and ensure that workers continue to play an important role in the industry. Second, the industry needs to rethink and redesign business models to ensure that both workers and companies benefit. Traditional business models may not meet the needs of modern technological development, so innovation and adjustment are necessary. For example, companies can increase productivity through the sharing economy model or flexible employment model, providing workers with more job opportunities and a more adaptable working environment. Additionally, enterprises should actively participate in developing and implementing fair labor policies to ensure that workers' working conditions and remuneration meet modern standards. Strong collaboration between businesses and education and training institutions is essential. This partnership can ensure that the education system aligns with industry needs, providing workers with practical skills and theoretical knowledge. In this way, enterprises can access a high-quality labor force while offering better career development opportunities for workers. To achieve this, substantial social policy reforms, such as

improvements in welfare and health protection systems, may be needed. These reforms should aim to provide workers with more comprehensive social security, including medical care, pensions, and unemployment benefits, ensuring that workers receive adequate support and protection when facing various work and life challenges. In conclusion, by adopting a people-centered and socially relevant approach, we can promote the equitable application of technological progress while protecting workers' rights. The industry needs to innovate business models, strengthen collaboration with education and training institutions, and promote social policy reforms to ensure that workers and companies benefit together. This approach will help build a more equitable, inclusive, and sustainable social and industrial environment.

(3) With the continuous development of technology, industry standardization and ecosystem construction are becoming increasingly important. These efforts promote the standardized application of technology and cross-industry cooperation, forming a more robust technological ecosystem. The formulation of standardization not only helps to solve compatibility and interoperability issues in current technology applications but also fosters the widespread adoption and in-depth application of technology, thereby promoting the healthy development of the entire industry. Industry standardization and ecosystem construction can be expanded in the following ways. Development of Unified Technical Standards and Protocols: This involves covering data formats, communication protocols, interface specifications, and privacy protection to ensure seamless integration and collaborative work between different systems and devices. Achieving this requires the participation and cooperation of industry associations, standardization organizations, enterprises, and research institutions. Establishment of a Cross-Industry and Cross-Platform Data Exchange and Sharing Mechanism: Promoting the opening and circulation of data fully taps into its value. By standardizing data interfaces and protocols, data interoperability between different systems and platforms can be achieved, improving data utilization efficiency and value creation. Construction of an Industrial Ecosystem with Digital Twin Technology at Its Core: This ecosystem should include technology providers, solution providers, application developers, end users, and other stakeholders. Through the coordinated development of the ecosystem, resources can be shared, advantages complemented, and the application and development of technology jointly promoted. Through these efforts, promoting the industry standardization and ecosystem construction of digital twin technology can effectively address various issues in technology application. This will improve the application level of technology, enhance industrial competitiveness, and promote the healthy and sustainable development of the entire industry.

In conclusion, while the Phygital Twin holds great promise for revolutionizing human–robot interaction, significant efforts are required to overcome the existing challenges and realize its full potential. Future research and development will need to bring together experts from robotics, computer science, cognitive science, and other related fields to drive innovation and progress in this exciting area.

References

1. Tao F, Zhang H, Liu A, Nee AYC (2019) Digital twin in industry: state-of-the-art. IEEE Trans Industr Inf 15(4):2405–2415
2. Jones D, Snider C, Nassehi A, Yon J, Hicks B (2020) Characterising the digital twin: a systematic literature review. CIRP J Manuf Sci Technol 29:36–52
3. Rasheed A, San O, Kvamsdal T (2020) Digital twin: values, challenges and enablers from a modeling perspective. IEEE Access 8:21980–22012
4. Aheleroff S, Xu X, Zhong RY, Lu Y (2021) Digital twin as a service (DTaaS) in Industry 4.0: an architecture reference model. Adv Eng Inform 47:101225
5. Wang B, Zhou H, Yang G, Li X, Yang H (2022) Human digital twin (HDT) driven human-cyber-physical systems: key technologies and applications. Chin J Mech Eng 35:11
6. Ortiz JN, Diaz PR, Sendra S et al (2020) A survey on 5G usage scenarios and traffic models. IEEE Commun Surv Tutorials 22(2):905–929
7. Kang B, Hwang I, Lee S et al (2018) My being to your place, your being to my place: co-present robotic avatars create illusion of living together. In: Proceedings of the 16th annual international conference on mobile systems, applications, and services, Germany, 10–15 June 2018, pp 54–67
8. Zhou H et al (2023) An attention-based deep learning approach for inertial motion recognition and estimation in human-robot collaboration. J Manuf Syst 67:97–110
9. Yang G, Lv H, Zhang Z et al (2020) Keep healthcare workers safe: application of teleoperated robot in isolation ward for COVID-19 prevention and control. Chin J Mech Eng 33:47
10. Selvaggio M, Cacace J, Pacchierotti C, Ruggiero F, Giordano PR (2022) A shared-control teleoperation architecture for nonprehensile object transportation. IEEE Trans Rob 38(1):569–583
11. Feizi N, Patel RV, Kermani MR, Atashzar SF (2022) Adaptive wave reconstruction through regulated-BMFLC for transparency-enhanced telerobotics over delayed networks. IEEE Trans Rob 38(5):2928–2942
12. Zhang J, Janeh O, Katzakis N et al (2019) Evaluation of proxemics in dynamic interaction with a mixed reality avatar robot. In: International conference on artificial reality and telexistence & eurographics symposium on virtual environments, Japan, 11–13 Sep 2019, pp 37–44
13. Beik-Mohammadi H, Kerzel M, Pleintinger B et al (2022) Model mediated teleoperation with a hand-arm exoskeleton in long time delays using reinforcement learning. In: 29th IEEE international conference on robot and human interactive communication, Italy, August 31–September 4, pp 713–720
14. Fang B, Wei X, Sun F et al (2019) Skill learning for human-robot interaction using wearable device. Tsinghua Sci Technol 24(6):654–662
15. Zhou H, Yang G, Lv H, Huang X, Yang H, Pang Z (2020) IoT-enabled dual-arm motion capture and mapping for telerobotics in home care. IEEE J Biomed Health Inform 24(6):1541–1549

GPSR Compliance

The European Union's (EU) General Product Safety Regulation (GPSR) is a set of rules that requires consumer products to be safe and our obligations to ensure this.

If you have any concerns about our products, you can contact us on ProductSafety@springernature.com

In case Publisher is established outside the EU, the EU authorized representative is:

Springer Nature Customer Service Center GmbH
Europaplatz 3
69115 Heidelberg, Germany

Batch number: 10704307

Printed by Printforce, the Netherlands